非洲电力可及现状分析

及可再生能源离网案例研究

水电水利规划设计总院 编著

·北京·

内 容 提 要

本书旨在梳理非洲电力可及现状的基础上，分析可再生能源离网项目开发模式的分类、判定和开发策略，为项目经济和可持续运营提供技术参考。同时，以中地海外尼日利亚阿布贾农业高科技产业园区为研究对象，通过开展案例研究，即结合中国"新能源＋农业"模式的实践经验，对该农业园区进行可再生能源离网升级改造的概念性设计，分析研究离网项目可发挥的经济效益和社会效益。

图书在版编目（CIP）数据

非洲电力可及现状分析及可再生能源离网案例研究：汉英对照 / 水电水利规划设计总院编著. -- 北京：中国水利水电出版社，2023.12
ISBN 978-7-5226-2162-3

Ⅰ. ①非… Ⅱ. ①水… Ⅲ. ①再生资源－发电－研究－非洲－汉、英 Ⅳ. ①TM61

中国国家版本馆CIP数据核字(2024)第018663号

责任编辑：殷海军　蔡晓洁

书　　名	**非洲电力可及现状分析及可再生能源离网案例研究** FEIZHOU DIANLI KEJI XIANZHUANG FENXI JI KEZAISHENG NENGYUAN LIWANG ANLI YANJIU
作　　者	水电水利规划设计总院　编著
出版发行	中国水利水电出版社 （北京市海淀区玉渊潭南路1号D座　100038） 网址：www.waterpub.com.cn E-mail: sales@mwr.gov.cn 电话：（010）68545888（营销中心）
经　　售	北京科水图书销售有限公司 电话：（010）68545874、63202643 全国各地新华书店和相关出版物销售网点
排　　版	中国水利水电出版社微机排版中心
印　　刷	北京印匠彩色印刷有限公司
规　　格	184mm×260mm　16开本　8.25印张　205千字
版　　次	2023年12月第1版　2023年12月第1次印刷
定　　价	**98.00元**

凡购买我社图书，如有缺页、倒页、脱页的，本社营销中心负责调换

版权所有·侵权必究

本书编委会

主　　任　李　昇　易跃春

副 主 任　顾洪宾　张益国　余　波　龚和平

主　　编　姜　昊　陈　长

编写人员　杨晓瑜　王宇亮　王先政　王　君　王沛元
　　　　　　周　港　帅　东　李彦洁　徐潇玉　夏玉聪
　　　　　　刘子初　靖赫然　黄　晋

咨　　询　苗　红　宋　婧　滕爱华　赵耀华　王　淼
　　　　　　汪　筠

序

电力是推动人类社会进步的重要物质基础,是当今社会和经济发展的关键因素。随着科技的发展和新技术的不断出现,用电需求持续增长,保障电力供应不仅是经济问题,更是关系国家能源安全、经济社会发展和民生福祉的社会问题。2021年,全球经济增长与更为极端的天气条件使全球电力需求增加了6%以上,全球各国加快电力供应,但无电地区人口的用电问题仍面临诸多挑战。

不断提升电力供应,实现100%的电力普及,是国际社会长期以来的奋斗目标。联合国可持续发展目标7提出:"到2030年,确保人人获得负担得起的、可靠和可持续的现代能源",确保不让任何一个人掉队。推动可再生能源发展,提升电力普及率植根于非盟《2063年议程》,是非洲各国领导人通过非盟采纳的长期愿景,并与非洲大陆对繁荣和包容性增长的期许高度一致。

截至2021年,全球仍有7.3亿无电人口,其中3/4生活在撒哈拉以南的非洲,非洲仍是全球电力可及问题最突出的地区。非洲人均能源消费量仅为180 kWh/a,远低于欧美国家每年上千甚至上万千瓦时的人均能源消费水平。据有关机构的研究显示,电力供应不足给非洲带来的经济损失约占其GDP的4%,并严重影响非洲人民的生活、生产、教育和医疗条件,已成为制约非洲经济和社会发展的重要掣肘。

推动提升非洲电力可及是国际社会共同关注的议题。早在2015年,G20会议就将应对气候变化、促进能源转型、改善撒哈拉以南非洲地区用电状况作为会议的重点议题,并写入《G20能源可及性行动计划:能源可及自愿合作》。中国长期以来同国际社会一道,致力于帮助非洲消除能源贫困,提升非洲电力可及水平。2021年10月,中国—非盟能源伙伴关系成立,同年11月,中非合作论坛第八届部长级会议通过的《达喀尔行动计划(2022—2024)》指出:"中方将同非方在中国—非盟能源伙伴关系框架下加强能源领域务实合作,共同提高非洲电气化水平，增加清洁能源比重，逐步解决能源可及性问题，推动双方实现能源可持续发展"。

随着应对气候变化和落实碳中和目标加速推进，世界各国正在大力推动能源转型和可再生能源发展。非洲的可再生能源资源禀赋优异，由于可再生能源具有清洁、低碳、可循环利用且可就地获取的特点，采用可再生能源离网发电系统可以在非洲偏远地区灵活布置。随着可再生能源离网技术不断成熟，建设成本不断下降，为解决非洲无电人口用电问题提供了新的思路。

Amb. Rahamtalla M. Osman
拉赫曼塔拉·奥斯曼
非洲联盟驻华代表处常驻代表

前　　言

能源是当今全世界共同关心的问题，处于几乎每一个主要挑战和机遇的核心。

——联合国

电力作为现代社会最为重要的能源之一，已经高度融入了人们生活的方方面面，推动提升电力可及是国际社会共同关注的议题。截至 2021 年全球约 7.3 亿无电人口中，3/4 生活在撒哈拉以南非洲，不难看出非洲仍是全球电力可及问题最突出的地区。中国国家能源局与非盟委员会于 2021 年签署谅解备忘录，发起建立中国—非盟能源伙伴关系，将逐步解决非洲电力可及问题作为双方务实合作的重要内容。

为什么非洲电力可及问题突出，且难以解决？非洲大部分无电人口都较为贫困，且分散居住在广阔的农村地区，地处偏远、交通阻塞，给电网建设带来了极大的施工不便和经济成本，因此传统的主网延伸方式难以有效解决非洲电力可及问题。近年来，随着可再生能源技术的不断进步，不依赖主网而采用可再生能源特别是光伏、风电等新能源独立供电运行的离网发电系统，为帮助消除非洲能源贫困，实现偏远地区人口电力普及提供了新的解决方案。

可再生能源具有清洁低碳、可循环利用、可就地获取、可在用户侧灵活布置等优势。因此，可再生能源离网发电系统既能有效解决非洲偏远地区无电人口用电问题，又能实现覆盖区域内的电力自给自足。值得注意的是，就非洲地区而言，无电人口多分布在偏远地区，当地居民的电价承受能力普遍较弱，仅依靠当地居民电费收入通常难以支撑离网项目经济和可持续运营。因此，为了促进可再生能源离网项目在非洲的落地和推广，还需要对开发模式进行深入研究。

本研究旨在梳理非洲电力可及现状的基础上，分析可再生能源离网项目开发模式的分类、判定和开发策略，为项目经济和可持续运营提供技术参考。同时，以中地海外尼日利亚阿布贾农业高科技产业园区为研究对象，通过开展案例研究，即结合中国"新能源＋农业"模式的实践经验，对该农业园区进行可再生能源离网升级改造的概念性设计，分析研究离网项目可发挥的经济效益和社会效益。

目　　录

序

前言

第 1 章　非洲电力可及现状分析 1

    1.1　电力可及是全球共同关心的重要议题 2

    1.2　非洲是全球电力可及问题最突出地区 6

    1.3　解决非洲电力可及问题全球正在行动 11

    1.4　推广非洲离网模式的重要意义 13

第 2 章　非洲离网开发模式分析 17

    2.1　离网项目用户群体的划分 18

    2.2　离网项目开发模式的分类 18

    2.3　离网项目开发模式的判断 19

    2.4　离网项目开发的策略 21

第 3 章　非洲离网项目案例研究 23

    3.1　项目概况——阿布贾农业园区 24

    3.2　农业园区离网改造概念性设计 31

    3.3　农业园区离网改造的综合效益 40

第 4 章　非洲离网项目推广探索 41

第 5 章　结论与建议 43

    5.1　结论 44

    5.2　建议 45

声明 47

图表目录

图 目 录

图 1-1　电力与人类生产生活的关系 …………………………………………… 2
图 1-2　2000—2030 年全球无电人口数量分布 ………………………………… 3
图 1-3　非洲各区域电力短缺情况 ……………………………………………… 10
图 2-1　开发模式判定流程图 …………………………………………………… 19
图 3-1　农业园区地理位置示意图 ……………………………………………… 24
图 3-2　农业园区发展定位 ……………………………………………………… 25
图 3-3　农业园区功能分区示意图 ……………………………………………… 25
图 3-4　农业园区用能示意图 …………………………………………………… 26
图 3-5　"新能源 + 农业"在非洲应用的 4 大优势 …………………………… 27
图 3-6　"新能源 + 农业"在中国常见的应用场景 …………………………… 27
图 3-7　光伏储能离网发电系统图 ……………………………………………… 32

表 目 录

表 1-1　非洲人均用电量与人均 GDP …………………………………………… 3
表 2-1　离网项目用户群体的划分 ……………………………………………… 18
表 2-2　离网项目开发模式的分类 ……………………………………………… 19
表 3-1　离网供电改造近期和远期情景 ………………………………………… 32
表 3-2　近期情景设备用电统计 ………………………………………………… 33
表 3-3　远期情景新增设备用电统计 …………………………………………… 33
表 3-4　光伏发电系统配置方案 ………………………………………………… 34
表 3-5　冷热电三联供系统配置方案 …………………………………………… 35
表 3-6　近期方案财务测算表 …………………………………………………… 38
表 3-7　远期方案财务测算表 …………………………………………………… 38
表 3-8　项目财务测算敏感性分析 ……………………………………………… 39
表 4-1　用电情景分类 …………………………………………………………… 42
表 4-2　三种情景下可再生能源离网投资测算 ………………………………… 42

缩略词

简称	英文全称	中文全称
STEPS	Stated Policies Scenario	既定政策情景
GOGLA	Global Off-Grid Lighting Association	全球太阳能离网协会
GDP	Gross Domestic Product	国内生产总值
SDSN	Sustainable Development Solutions Network	可持续发展解决方案网络
SDG7	Sustainable Development Goal 7	可持续发展目标 7
ESMAP	Energy Sector Management Assistance Program	能源部门管理援助计划
PPIAF	Public Private Infrastructure Advisory Facility	公私基础设施咨询机构
IMELS	Italian Ministry for the Environment Land and Sea	意大利环境、领土与海洋部
ROGEP	Regional Off-Grid Electrification Project	区域离网电气化项目
IDA	International Development Association	国际开发协会
SEFA	Sustainable Energy Fund for Africa	非洲可持续能源基金
CRP	Covid-19 Off-Grid Recovery Platform	离网恢复平台项目
IEA	International Energy Agency	国际能源署
REA	Rural Electrification Agency	农村电气化局
OGES	Off-grid Electrification Strategy	离网电气化策略
PSRP	Power Sector Recovery Programme	电力行业复苏计划
KOSAP	Kenya Off-Grid Solar Access Project	肯尼亚离网光伏项目
KNES	Kenya National Electrification Strategy	肯尼亚国家电气化战略
REMP	Reneable Energy Master Plan	可再生能源总体规划
SREP	Scaling-up Renewable Energy Programme	扩大可再生能源项目
ONEE	National Office of Electricity and Drinking Water	国家水利电力总局
IRENA	International Renewable Energy Agency	国际可再生能源机构
ROGEA	Regulation for Off-Grid Energy Access	离网能源可及条例
MEFA	Mozambique Energy for All	莫桑比克人人享有能源
FUNAE	Mozambique National Energy Fund	莫桑比克国家能源基金
SHS	solar home system	太阳能家庭系统
NESPA	Niger Solar Electricity Access Project	尼日尔太阳能电力可及项目
NST1	National Strategy for Transformation	国家转型战略
AFDB	African Development Bank	非洲开发银行
LCOE	Levelized Cost of Energy	平准化度电成本

第 1 章
非洲电力可及现状分析

非洲拥有巨大的经济发展红利，但电力可及问题长期制约非洲可持续发展。本章分析了在国际社会高度重视电力可及的背景下，解决非洲无电人口用电问题的重要性，梳理了非洲电力可及现状与挑战，论述了可再生能源离网模式对于解决非洲电力可及问题的重要现实作用。

背景

目标 7.1：到 2030 年，确保人人都能获得负担得起的、可靠和可持续的现代能源。
指标 7.1.1：城市/农村用电人口比例 (%)。

—— 联合国可持续发展目标 7

图 1-1 电力与人类生产生活的关系

联合国把确保人人都能获得负担得起的、可靠和可持续的现代能源作为人类社会可持续发展的重要目标之一，并将能获得电力的人口比例作为衡量该目标实现程度的重要指标。电能的消费水平，即电力可及程度已成为表征一个国家现代化进程和人民生活水平的重要标志。很难想象，如果没有电力等现代能源服务，今天的世界将如何运作。

电力是现代社会和经济运行的动力之源，是支撑人类文明进步和经济社会发展的基础性能源资源。电力的发展不仅关系到国家经济安全等重大战略问题，还与人们的日常生活、社会稳定密切相关。从农业、工业和生产到医疗、教育和通信等各个领域，电力几乎已经成为现代社会各领域不可或缺的重要保障（图 1-1）。实现电力可及是缓解贫困、提高社会生产力和促进共同繁荣的必要先决条件。

1.1 电力可及是全球共同关心的重要议题

解决电力可及和应对气候变化问题是各国面临的双重任务。

—— 联合国秘书长古特雷斯

截至 2021 年，全球仍有 7.3 亿人在没有电力供应的情况下生活，约占全球人口的 10%。根据目前的发展趋势，距离实现 2030 年联合国可持续发展目标 7 仍面临严峻挑战，全球仍有许多地区没有电力接入，妇女和儿童必须花费几个小时来取水，诊所不能储存儿童疫苗，许多学生晚上不能做家庭作业。若国际社会不能立即采取行动，预计到 2030 年全球仍将有 6.5 亿人无法获得电力供应（图 1-2）。这意味着，联合国此前提出的确保在 2030 年人人获得负担得起的、可靠和可持续的现代能源的目标恐难以实现。当前，进一步提升可再生能源利用比例是实现电力可及的关键。

图 1-2 2000—2030 年全球无电人口数量分布

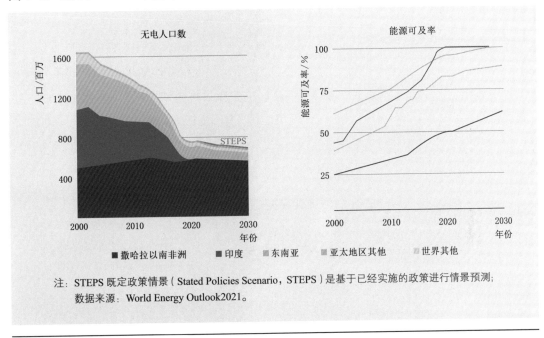

非洲地区共有 54 个主权国家，高用电区域主要集中在北部非洲和南部非洲，这两个地区的用电量占全非洲用电量的 80% 左右，人均用电量约为 2000 kWh/a。其他非洲地区的人均用电量约为 200 kWh/a，不足欧洲人均用电量的 1/30，不足美国人均用电量的 1/60，该地区的许多非洲国家人均用电量甚至低于 100 kWh/a。非洲 54 个国家的人均用电量与人均 GDP 统计见表 1-1。可以看出，电力可及率低于 50% 的非洲国家有 28 个，占到了全部主权国家的 1/2 以上，这些国家全部分布在撒哈拉沙漠以南地区，电力供应不足在很大程度上影响了当地的经济发展水平。

表 1-1 非洲人均用电量与人均 GDP

国　　家		人均用电量/(kWh/a)	电力可及率 /%	人均 GDP/(美元 /a)
北非地区	阿尔及利亚	1302	99	3263
	埃及	1534	99	3587
	利比亚	3962	99	3281

续表

国家		人均用电量/(kWh/a)	电力可及率/%	人均GDP/(美元/a)
北非地区	摩洛哥	794	99	3158
	突尼斯	1303	99	3323
中非地区	喀麦隆	231	63	1470
	中非共和国	27	5	490
	乍得	12	8	654
	刚果	172	48	2186
	刚果民主共和国	73	9	541
	赤道几内亚	556	67	6773
	加蓬	928	91	7421
东非地区	布隆迪	32	10	254
	吉布提	409	42	3074
	厄立特里亚	58	50	588
	埃塞俄比亚	84	48	994
	肯尼亚	147	78	2039
	卢旺达	41	55	819
	索马里	27	35	327
	南苏丹	34	7	296
	苏丹	266	47	775
	乌干达	72	26	912
西非地区	尼日利亚	115	68.2	2083
	贝宁	89	31	1251
	科特迪瓦	274	78	2278
	加纳	319	85	2223
	塞内加尔	222	70	1460
	多哥	146	46	905
	布基纳法索	74	21	791
	佛得角	630	95	3148
	冈比亚	130	62	791

续表

国家		人均用电量 / (kWh/a)	电力可及率 /%	人均 GDP/ (美元 /a)
西非地区	几内亚	44	44	1106
	几内亚比绍	19	39	790
	利比亚	55	29	3281
	马里	153	52	897
	毛里塔尼亚	264	47	1971
	尼日尔	47	14	566
	圣多美和普林西比	291	77	1918
	塞拉利昂	42	22	527
南非地区	安哥拉	278	45	2012
	博茨瓦纳	1569	60	6781
	科摩罗	46	70	1362
	莱索托	430	46	1003
	马达加斯加	59	27	502
	马拉维	62	11	407
	毛里求斯	1976	99	8993
	莫桑比克	384	38	450
	纳米比亚	1479	45	4175
	塞舌尔	3391	99	11639
	斯威士兰	1296	76	3054
	坦桑尼亚	97	38	1090
	赞比亚	634	33	981
	津巴布韦	489	53	1385
	南非	4064	99	5059

电力普及为什么刻不容缓？

♦ 根据洛克菲勒基金会的数据，电力的接入可以使家庭人均收入增加39%，企业以更高的生产水平运作，农民可运行清洁的灌溉系统和加工机器，提高产量，增加收入。

♦ 摩洛哥一个名为伊姆加迪（Id Mjahdi）的偏远山村，自通过太阳能实现通电以来，儿童可以在夜间读书，人们不用再到几公里外的水井取水，还为当地带来了制作坚果油等产品的就业机会，增加了当地家庭收入。

◆ 全球性离网太阳能协会(GOGLA)在东非地区调研了2300户家庭，有约58%的家庭安装了离网太阳能系统，约36%的家庭因此月收入平均增加了35美元，相当于该地区家庭月平均GDP的50%。

◆ 撒哈拉以南非洲的部分国家因得不到可靠的电力供应，国内生产总值每年损失的比例可能接近7%，30%的医疗设施场所缺乏电力，这导致了2.55亿人口医疗条件受损。

◆ 目前地球上每10个人当中约有1个人用不上电，因此得不到电力带来的工作、学习或经营的机会，也无法享受更加安全、有保障的医疗服务。

◆ 由于无法获得电力和清洁燃料，全球仍有约24亿人使用木柴、煤炭、粪便等方式取暖和做饭，使许多家庭饱受烟熏火燎之苦，并且患上心脏病、中风、癌症、肺炎等疾病，每年因此有数百万人死亡。

1.2 非洲是全球电力可及问题最突出地区

非洲不能在黑暗中发展。

——非洲联盟副主席奎西·夸第

非洲许多儿童还无法使用
电灯进行夜间阅读

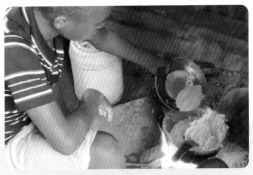
非洲许多家庭仍无法使用
电力进行清洁烹饪

非洲许多商铺还没有配备
必要的电气化设备

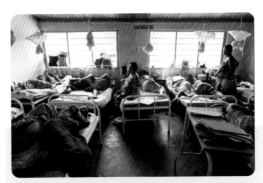
非洲许多医院仍非常缺乏
电气化的医疗设施

非洲的电力可及问题十分突出。全球有约 7.3 亿无电人口，其中 3/4 生活在撒哈拉以南的非洲，该地区的许多非洲国家在未来较长一段时间内仍无法获得满足用能需求的电力基础设施。截至 2021 年，非洲大陆约有 6.3 亿人仍缺乏电力供应，占非洲人口总数的 42%。

目前，非洲国家人均用电量约为 200 kWh/a，不足一台冰箱的年耗电量，与欧洲每年 6500 kWh 和美国每年 13000 kWh 的人均用电量相比显得微不足道。基础设施匮乏，电力投资不足，电网覆盖率低，新能源开发率低，人口过于分散，以及技术和管理水平较为薄弱等问题是导致电力普及率低的主要原因，当前电力供应不足已成为制约非洲经济和社会发展的关键掣肘。

非洲拥有巨大的经济社会发展红利

非洲是世界第二大洲，在全球舞台上有举足轻重的地位。这是一片蕴含无限可能的热土，幅员辽阔，土地、能源得天独厚，在世界政治经济格局中的战略地位十分重要。近20年来，非洲经济持续增长，并预期将保持成为全球发展最快的地区之一。

- 非洲土地广阔肥沃，总面积为3020万 km^2，可利用的土地资源广阔，75%的面积都是高原和平原。
- 非洲人口红利巨大，总人口数为12.8亿，仅次于亚洲，位居世界第二，50%以上是20岁以下的青年人口。
- 非洲的清洁能源资源丰富，拥有世界第一长河尼罗河及世界第二大水系刚果河，并且风能和太阳能资源丰富，水能、风能、太阳能资源分别占全球的12%、32%和40%。
- 世界银行承诺50亿美元用于6个非洲国家能源项目的新技术和财政支持。中国政府把对非电力投资作为"一带一路"建设一个重要且优先的合作议题。

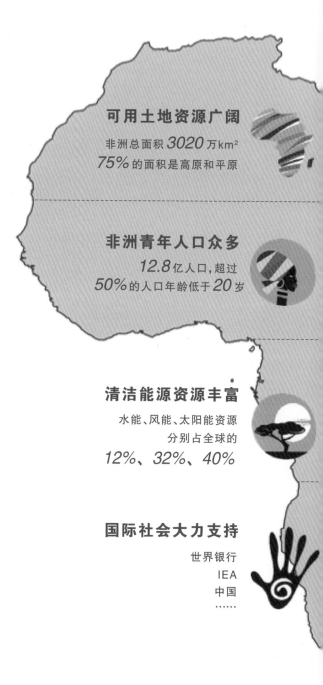

可用土地资源广阔
非洲总面积 *3020* km²
75% 的面积是高原和平原

非洲青年人口众多
12.8 亿人口，超过 *50%* 的人口年龄低于 *20* 岁

清洁能源资源丰富
水能、风能、太阳能资源分别占全球的
12%、*32%*、*40%*

国际社会大力支持
世界银行
IEA
中国
……

电力可及问题制约非洲可持续发展

增加经济成本
因电力短缺造成的经济损失占非洲GDP的 *4%*

制约工业生产
频繁停电带来的损失占到企业平均销售额的 *6%*

影响福祉民生
供电不足严重影响非洲人民的生活、生产、教育和医疗条件

带来环境污染
约 *9.7* 亿非洲人在使用木柴、煤、木炭或动物粪便作为主要生活能源

撒哈拉以南非洲地区人口占世界的13%，却只占世界能源消费量的4%，人均能源消费量仅为世界平均水平的1/3，发电装机容量仅占全球不到3%。大量的无电人口（占世界无电人口的75%）与电力短缺依然是制约非洲社会经济发展的首要问题，低水平收入以及低效、昂贵的能源供应形式使得能源负担能力成为关键挑战。

◆ 30多个非洲国家正饱受着电力短缺和经常性服务中断：平均每年塞内加尔25天、坦桑尼亚63天、布隆迪144天……

◆ 尼日利亚作为非洲最大的石油输出国组织成员国，尼日利亚企业却因为电力供应不稳定，每年约损失290亿美元。

◆ 2022年7月，南非全国正在执行该国历史上第二次六级限电，每日约6个小时以上的停电已经导致南非民众正常的生产生活受到严重影响。

◆ 非洲薪柴、木炭等的使用向空气中排放了大量污染物和二氧化碳，非洲因空气污染每年造成的经济损失高达数千亿美元。

非洲无电人口约 6.3 亿，主要集中在撒哈拉以南的地区。非洲总体的通电率为 58%，各区域的通电率差异较大，呈现北高南低的特点，北非地区电力普及程度很高，通电率已达到 98%，其他区域距离实现电力可及的目标还有较大差距。因此，为帮助非洲加快实现电力可及进度，应重点为撒哈拉以南的非洲地区提供可靠和广泛的电力供应。

图 1-3 非洲各区域电力短缺情况

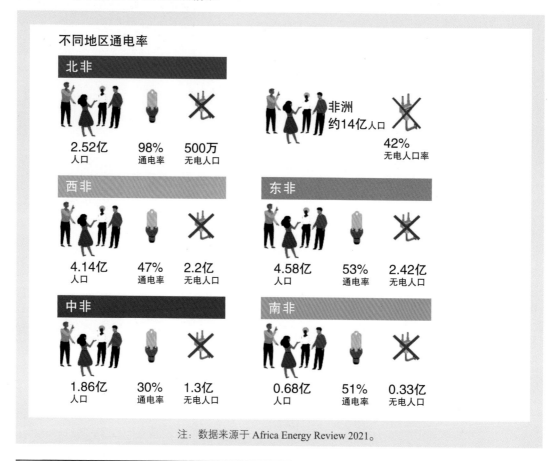

注：数据来源于 Africa Energy Review 2021。

- 国际能源署预测到 2030 年，世界上仍有约 8% 的人口用不上电力，即约 6.7 亿人得不到电力服务。需要加快推动非洲等欠发达地区国家、脆弱和受冲突影响国家实现电力可及的进程。

- 非洲的通电率从 2010 年的 49% 提高到 2020 年的 58%，电力可及取得积极进展。但是值得注意的是，这 10 年间实现电力可及的人口主要集中在北非地区，相比之下撒哈拉以南非洲地区电力可及率上升较为缓慢，这使得最弱势人群更加落后。

- 国际能源署发布的《2022 年非洲能源展望》显示，受新冠疫情的影响，2020 年非洲无电人口数量增加了 1300 万，这是非洲自 2013 年以来首次偏离了为 6 亿人消除能源贫困的目标，结束了过去 6 年间电力可及的进步趋势。
- 可持续发展解决方案网络（SDSN）研究显示，54 个非洲国家实现联合国可持续发展目标 7（SDG7）均面临挑战，超过 75% 的国家仍面临"重大挑战"。其中，只有加蓬有望实现 SDG7，3 个国家"下降"，28 个国家"停滞"，其余 22 个国家仅"适度增长"。
- 调动金融机构和捐助者的积极性，对于减少非洲区域发展不平衡的差距，推进非洲电力可及进程至关重要。据国际能源署的研究显示，为了实现联合国 2030 年电力可及目标，按照目前的进度，每年需要投入约 250 亿美元，约相当于疫情前非洲能源总投资的 1/4。

1.3 解决非洲电力可及问题全球正在行动

非洲和国际社会当务之急是为所有非洲人提供且负担得起的现代能源。到 2030 年实现这一目标，可以通过每年投资 250 亿美元来实现。在我们力所能及的情况下，非洲能源贫困的持续不公正问题却得不到解决，这在道义上是不可接受的。

——国际能源署执行董事 Fatih Birol

推动非洲电力可及工作是国际社会的长期愿景。为了解决经济和社会发展长期受制于电力匮乏的难题，务实提高人民生活福祉，世界各国一直致力于加快非洲可再生能源发展，希望利用当地丰富的可再生能源资源，通过建设能源基础设施，确保所有非洲家庭、工厂和企业等都能拥有现代、高效、可靠、经济、低碳的清洁能源，创造更多就业机会，推动经济社会全面发展。

世界银行

世界银行于 2007 年发起"点亮非洲"项目，是为实现"人人享有可持续能源"作出的重要贡献之一。其主要面向非洲国家，计划到 2030 年使撒哈拉以南非洲 2.5 亿无电人口获得清洁、负担得起、质量经过验证的离网照明和能源产品。"点亮非洲"项目得到了能源部门管理援助计划 (ESMAP)、公私基础设施咨询基金 (PPIAF)、荷兰外交部、意大利环境、土地和海洋部 (IMELS) 和宜家基金会的支持。目前已有超过 3200 万非洲人民通过该项目实现了电力可及。

世界银行于 2019 年批准了区域离网电气化项目 (ROGEP)，其中包括来自国际开发协会 (IDA) 的 1.5 亿美元信贷和赠款，以及来自清洁技术基金的 7470 万美元应急恢复赠款，以帮助西非 19 个国家和萨赫勒地区离网供电。该项目旨在使用区域协调的方法为独立的太阳能系统创建区域市场,以增加区域内的家庭、企业和公共机构的电力供应,预计惠及约 170 万无电人口。

非洲开发银行

非洲开发银行(AFDB)于 2016 年启动《非洲能源新政战略》。截至 2020 年,投入 120 亿美元资金,并调动 8.5 亿美元融资,以促进能源普及和清洁能源转型。该战略鼓励采用分布式能源,旨在通过太阳能家庭系统等离网技术帮助 300 万人获得电力。2020 年肯尼亚获得非洲开发银行约 1.5 亿美元资金,实施肯尼亚离网电气化计划,为 14 个县的 25 万户家庭提供太阳能。

2021 年 8 月,非洲开发银行已经就非洲可持续能源基金(SEFA)为 Covid-19 电网外恢复平台(CRP)提供 2000 万美元优惠投资的融资协议达成了共识。该平台支持企业将太阳能家庭系统、绿色微型电网、清洁烹饪和其他可再生能源电力可及解决方案的商业化,以减轻疫情影响并推动行业复苏。这项为期 5 年、总价 5000 万美元的融资倡议旨在为能源可及领域提供救济和恢复资金。

G20

20 国集团(G20)是全球最大的经济组织之一,经济总量占世界的 90%,长期以来把能源可及作为能源领域的重要议题。早在 2015 年,G20 成员国通过了《G20 能源可及性行动计划:能源可及自愿合作》,根据各国实际情况和发展要点,自愿承诺加强合作和知识经验分享,共同为全球 10% 无电人口提供电力。第一阶段的工作重点是改善撒哈拉以南非洲地区的用电状况。

由于新冠疫情导致全球无电人口数量首次增加,2020 年 9 月的 G20 能源部长会议发布公报,重申了关于加快能源可及的承诺,并呼吁成员国和相关国际组织再次在自愿的基础上考虑加快清洁烹饪和电气化进程,强调构建以结果为导向的融资机制;加强清洁烹饪市场建设工作;协助政府制定需要各国承诺的综合能源规划;加强目标国家公共和私营部门能力建设等。

中国

在"一带一路"倡议的带动下,中国企业在非洲约 70% 的国家开展电力合作。2010—2020 年,中国企业参与了非洲约 150 个电厂和输配电项目建设,非洲超过 1 亿人口通过接入电网获得电力,其中,中国企业的贡献率达到 30%。IEA 于 2019 年预测,中国到 2024 年前,预计完工 49 个在非洲承建的发电项目,其中绝大部分是可再生能源项目,相当于该地区同期装机总量的 20%。

中国国家能源局和非盟委员会于 2021 年签署谅解备忘录,同意成立中国—非盟能源伙伴关系,并将提升非洲电力可及作为合作的重要内容。中非合作论坛第八届部长级会议通过的《达喀尔行动计划(2022—2024)》指出:"中方将同非方在中国—非盟能源伙伴关系框架下加强能源领域务实合作,共同提高非洲电气化水平,增加清洁能源比重,逐步解决能源可及性问题。"

1.4 推广非洲离网模式的重要意义

非洲用电问题难以解决的一个重要因素是非洲的无电人口 80% 以上都是农村人口,地处偏远、交通阻塞、人口分散,而且非洲地区的电网覆盖较为薄弱,很多地区甚至还没有覆盖电网,因此依靠主网延伸的方式解决这部分人的用电问题,总投资太大。非洲的可再生能源可就地获取、离网布置灵活且技术成熟,并且离网相比于主网,投资建设的总成本低很多,符合非洲的实际情况。因此,国际社会日益重视采用可再生能源离网模式来解决无电人口的用电问题。

什么是离网?

离网是一种采用区域独立发电、分户独立发电的供电模式,不依赖主干电网而独立运行的发电系统。这种发电系统由于不受地域限制,可以在不同条件的地区广泛使用,多采用可再生能源进行供电,因此非常适合于偏远无电网地区,也可用作经常停电地区的应急发电设备。对于无电网地区或经常停电地区的家庭、企业和工厂等用户来说,具有很强的实用性,且采用可再生能源离网相比于柴油发电经济性更优。因此,离网发电系统对于无电网地区或经常停电地区具有很好的应用潜力。

离网有哪些特点？

总投资低

离网系统布置灵活，可在用户侧就近安装，能充分利用当地可再生能源资源，无需建设大型电站及远距离输电网，总成本较低。

建设较快

离网的建设周期短见效快，易复制性强，便于集中管理，受到空间的限制小，可以根据需求增长来灵活调整容量，极易组合和扩容。

环保低碳

离网光伏发电碳排放量是化石能源发电的 1/20～1/10，采用可再生能源作为主力电源来实现自发自用，是绿色低碳的发电形式。

融资性强

当今国际社会非常支持小而美的离网项目，世界银行、非洲开发银行以及许多非洲国家都发起了离网项目的融资激励措施。

惠及民生

离网系统因其可配合新能源在各地灵活布置、安装简便、用途广泛，可满足远离大电网地区的工业、农业、商业、医学等各领域用电需求。

推动非洲离网有哪些举措？

非洲可再生能源离网模式推广潜力巨大，对于满足非洲社会用电需求、解决无电人口用电问题，以及推动能源转型具有重要的现实作用。非洲各国正在积极制定国家电气化战略，纷纷将离网发展纳入其国家电气化战略的重要组成部分。部分非洲国家为推动可再生能源离网模式所采取的激励政策如下。

1. 尼日利亚

尼日利亚农村电气化局（REA）过去专注于电网扩展投资，但现在专注于为农村居民提供离网解决方案。由 REA 制定的离网电气化战略（OGES），作为电力行业复苏计划（PSRP）的组成部分，旨在为家庭、社区和企业供电提供分布式能源解决方案，计划到 2023 年建成 10000 个微型电网，为 14% 的人口提供电力；为 25 万家中小企业提供可靠的电力供应；为联邦大学和大学教学医院提供不间断的电力供应；为住宅和中小企业部署 500 万套独立太阳能系统。

2. 埃塞俄比亚

埃塞俄比亚政府早在 2016 年就制定了农村电气化（REF）战略，旨在实现农村离网地区供电，分 5 期完成 45365 个太阳能家庭系统项目，帮助约 545 个农村卫生站和约 370 所小学和培训中心通电。埃塞俄比亚政府于 2019 年制定了国家电气化计划 2.0（NEP2.0），目标是到 2025 年实现全国 100% 电力普及，其中包括通过离网为国家 35% 的人口提供电力。随着该计划的实施，到 2025 年全国将通过独立的太阳能解决方案和微型电网技术，新增 600 万个离网接入。

3. 肯尼亚

肯尼亚离网光伏项目 (KOSAP) 是肯尼亚能源部的旗舰项目，于 2017 年开始实施，旨在为该国偏远、低密度和传统上服务不足的地区提供电力和清洁烹饪解决方案。该项目是肯尼亚国家电气化战略（KNES）的重要组成，也是肯尼亚为实现 2030 年愿景的重要激励措施，该项目包括家庭和公共设施的微型电网、公共设施独立家庭系统、社区设施用太阳能水泵、能力建设提升等内容。随着该项目的实施，肯尼亚将建成 151 个微型电网，并提供 25 万个独立太阳能家庭系统。

4. 加纳

加纳政府于 2019 年发布了可再生能源总体规划（REMP），将离网电气化项目纳入该规划，计划为 1000 个离网社区提供可再生能源离网发电方案。近年来，加纳政府正在大力推进扩大可再生能源计划（SREP），旨在通过释放融资机会，加快可再生能源发展。其中，离网项目包括 55 个政府投资的可再生微型电网，以及由私营部门为 33000 户家庭、1350 所学校、500 个医疗中心和 400 个社区投资的独立太阳能系统。该计划于 2021 年完成相关准备工作，于 2022 年开始实施。

5. 摩洛哥

摩洛哥自 2016 年签署《巴黎协定》以来，加快推动能源转型和可再生能源发展，计划到 2030 年实现由可再生能源满足 50% 的电力需求，到 2050 年实现由可再生能源满足 100% 的电力需求。可再生能源离网是该战略的重要内容，2018 年，摩洛哥国家水利电力总局（ONEE）通过与第三方合作，为摩洛哥 1000 多个村庄的 19438 个家庭安装了离网太阳能系统。根据 IRENA 数据，摩洛哥全国现有约 128000 户家庭通过太阳能家庭系统供电，是非洲采用该技术排名前 3 位的国家。

6. 莫桑比克

莫桑比克政府于 2021 年 12 月批准了离网能源可及条例（ROGEA），为离网领域提供法律框架支持，增加相关参与者的透明度，为私营部门提供必要条件，并保护各类离网技术的投资，如太阳能家庭系统和微型电网等。推动离网电气化是 2021 年莫桑比克人人享有能源项目（MEFA）的重要组成部分，计划到 2030 年，实现全国 100% 的电力普及，其中，30% 的电力将由离网发电提供。莫桑比克国家能源基金（FUNAE）于 2022 年 6 月获得了世界银行 2600 万美元资金的支持，以促进离网发展，并推广太阳能家庭系统（SHS）项目，计划在未来 4 年内惠及约 30 万人。

7. 尼日尔

尼日尔政府于 2017 年与世界银行国际开发协会 (IDA) 达成协议，IDA 计划提供 5030 万美元，用于支持尼日尔太阳能电力可及项目 (NESPA)，旨在通过太阳能实现尼日尔农村地区通电。NESPA 的主要工作内容包括独立光伏设备的市场开发、太阳能微网实现农村电气化、孤立式太阳能混合微网系统、执行支持和技术援助等，该项目将惠及尼日尔 352 个村庄，推动尼日尔农村地区电气化。

8. 卢旺达

根据卢旺达政府制定的 7 年政府计划：国家转型战略（NST1，2017—2024），把离网供电作为该国实现电气化的重要内容，提出了发电、供电质量和可靠性以及供电能力等方面的发展目标，计划到 2024 年以最快和最经济的方式实现全国 100% 电力可及，其中，主网供电比例为 52%，离网供电比例为 48%。卢旺达的离网技术近年来逐渐成熟，正在为实现 100% 电力可及的目标发挥着重要作用，截至 2022 年 6 月，卢旺达离网供电比例已经达到 22%。

第 2 章
非洲离网开发模式分析

本章以离网项目可持续运营为目标导向，按照是否需要外部资金补贴以及补贴方式对开发模式进行分类，并通过比较用户侧电价承受能力与项目平准化度电成本，提出离网开发模式的选择方法，在保障低用电群体享有普惠性用电权益的基础上，实现离网可持续运营。

离网项目的受众多分布在偏远地区，非洲偏远地区经济基础薄弱，项目所在地的用户电价承受能力通常较弱，仅依靠当地居民的电费收入难以实现离网项目经济可持续运营，需要对离网项目的开发模式进行分析。

2.1 离网项目用户群体的划分

为了保障离网项目经济可持续运营，应对项目的用户群体进行分类。通过不同用户群体的合理配置，一方面在一定程度上保障项目的经济性；另一方面在公平、合理的原则下进行电价分摊，通过高收入群体对低收入群体的交叉补贴，保障低收入群体享有普惠性的基本用电权益。

按照离网项目的用电需求和支付能力，离网项目用户群体的划分见表2-1，将用户群体划分为三类，分别是第一类用户、第二类用户、第三类用户。第一类用户是指具有较高用电需求且支付能力稳定的用户，可以在较大程度上保障项目经济可持续运营，一般为大中型工业用户，例如：一定规模的矿场、冶金和生产车间等。第二类用户是指具有中等电力需求且支付能力较稳定的用户，可以在一定程度上保障项目经济可持续运营，一般为普通工商业用户，例如：集市、菜场、医院、学校、商业楼等。第三类用户是指具有较小用电需求且支付能力欠稳定的用户，只能作为离网项目的电费补充，无法保障项目经济可持续运营，例如：普通家庭和个体户等。

表2-1 离网项目用户群体的划分

用户类别	用电需求	支付能力	主要作用	用户组成
第一类	较高	稳定	较大程度上保障项目经济性	矿场、冶金、生产车间等
第二类	中等	较稳定	一定程度上保障项目经济性	集市、菜场、医院、学校、商业楼等
第三类	较小	欠稳定	电费补充，无法保障项目经济性	普通家庭、个体户等

2.2 离网项目开发模式的分类

离网项目开发模式的分类需要根据当地的实际情况进行科学判断。由于不同地区各类群体分布的差异性，电费收入不一定能够完全满足项目纯商业化运营的要求，部分项目的可持续运营需要依靠政府或其他外部补贴。根据是否需要外部补贴以及补贴方式的不同，离网项目的开发模式可以分为盈利模式、部分补贴模式和全额补贴模式三类（表2-2）。

表 2-2　离网项目开发模式的分类

开发模式	目的	补贴方式	模式特点
盈利模式	盈利	无补贴	电费收入能够全面覆盖项目前期的资本成本（含权益成本）和后期运维成本，无须外部补贴即可商业运营
部分补贴模式	盈利+扶贫	补贴前期资本成本	一次性对项目前期的资本成本进行外部补贴，后期通过电费收入覆盖运维成本，实现项目可持续运营
全额补贴模式	扶贫	全过程补贴	电费收入无法覆盖后期的运维成本，须全过程外部补贴实现项目可持续运营

盈利模式遵循市场商业原则，电费收入能够全面覆盖项目前期的资本成本（含权益成本）和后期运维成本；部分补贴模式不以盈利为首要目的，兼具盈利和扶贫双重属性，电费收入可以覆盖项目后期的运维成本，但是项目前期的资本成本（含权益成本）全部或大部分需要依靠政府或其他外部补贴；全额补贴模式不以盈利为目的，具有扶贫属性，电费收入既无法覆盖项目前期的资本成本（含权益成本），也无法覆盖项目后期的运维成本，项目建设和运行需要完全依靠政府或外部提供的赠款、补助和优惠贷款。

2.3　离网项目开发模式的判断

根据表 2-2，离网项目三类开发模式的选择判断，与电费收入在多大程度上覆盖项目全生命周期成本（含权益成本）密切相关。而电费收入取决于用户侧综合可承受电价（$\overline{C}$），项目全生命周期成本（含权益成本）可由平准化度电成本（$LCOE$）直观反映。现将离网项目三类开发模式的判断方法介绍如下：

（1）当 $\overline{C} \geqslant LCOE$ 时，不需要外部补贴，项目为盈利模式。

（2）当 $\overline{C} < LCOE$ 时，若 $P_{O\&M} \leqslant \overline{C} < P_I$，电费收入可以覆盖项目后期的运维成本，

图 2-1　开发模式判定流程图

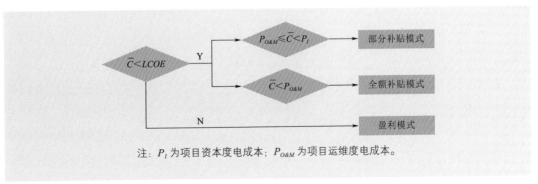

注：P_I 为项目资本度电成本；$P_{O\&M}$ 为项目运维度电成本。

但无法覆盖项目前期的资本成本（含权益成本），项目采用部分补贴模式。

（3）当 $\overline{C} < LCOE$ 时，若 $\overline{C} < P_{O\&M}$，电费收入不仅无法覆盖项目前期的资本成本（含权益成本），且无法覆盖项目后期的运维成本，项目采用全额补贴模式。

专栏一：平准化度电成本

平准化度电成本（Levelized Cost of Energy，LCOE）作为一个量化的经济指标，常被用于比较和评估能源发电项目的综合经济效益，其是对项目生命周期内的成本和发电量进行平均化后计算得到的发电成本，即项目全生命周期内的成本现值与全生命周期内发电量现值的比值。平准化度电成本在国际上被广泛地使用，用以衡量不同能源发电项目收益和成本的关系，并为售电电价和售电方案的制定提供参考。

平准化度电成本的计算公式（IRENA，2018）为

$$LCOE = \frac{\sum_{t=1}^{n} \frac{I_t + M_t + F_t}{(1+r)^t}}{\sum_{t=1}^{n} \frac{E_t}{(1+r)^t}} \quad (2\text{-}1)$$

$$M_t = f(dep, fin, O\&M, ret)$$

其中

式中 $LCOE$ ——平准化度电成本；

I_t ——第 t 年投资（含投资回报）；

M_t ——第 t 年经营成本；

dep ——折旧费；

fin ——财务费用；

$O\&M$ ——运维费用；

ret ——售电费用；

F_t ——第 t 年燃料费；

E_t ——第 t 年发电量；

r ——贴现率；

n ——财务计算周期。

专栏二：用户综合可承受电价

用户综合可承受电价决定了用户侧各类用户可以承受的平均电价上限，是科学、合理制定电价政策的基础。通常情形下，用户综合可承受电价需通过现场调研确定，通过对用户侧各类典型用户的人均可支配收入、电费占可支配收入比例和用电量等数据进行分析、计算确定。

2.4 离网项目开发的策略

离网项目的开发需要从当地的实际用能需求出发，根据用户群体的电价承受能力和项目的实际情况，科学定位项目开发模式。为了在尽可能减少政府财政补贴的情况下促进离网项目开发，从而保障市场稳步和可持续发展，不断惠及偏远地区无电人口，政府主管部门应该本着"科学统筹、因地制宜、循序渐进"的原则，据实针对三类项目制定不同的项目开发策略，并出台相应的激励措施。建议大力推进盈利项目，优先遵循市场商业化原则解决无电人口用电问题；适当支持部分补贴项目，通过一定程度的政策扶持实现项目自我造血和可持续运营；择优开展全额补贴项目，对于民众需求迫切、示范意义突出、普惠作用明显，但又不具备商业运营条件的民生项目，考虑采用非市场化的扶贫方式给予支持。

离网项目开发策略

大力推进盈利项目

为了保障离网市场健康发展，应该尽可能减少政府干预和财政支出，优先遵循市场化方式实现项目可持续运营，因此在开发策略上建议大力推进盈利项目。在项目规划时应重点考虑区域内是否有较高电力需求且支付能力稳定的第一类用户，优先布局在工矿企业等较高负荷用能区、集中的优质产业区以及政策有利区附近，从根本上保障离网项目的经济性。

适当支持部分补贴项目

为了惠及更多民生，满足电价承受能力较低用户的用电需求，政府应该对离网项目前期建设进行一定的财政补贴，使项目后期实现自我造血，因此建议适当支持部分补贴项目。在项目规划时应重点考虑区域内是否有一定规模的第二类用户，以及电费是否可以覆盖后期运营成本，优先布局在从事生产经营活动且用电较为集中的商户附近，促进区域经济均衡发展。

择优开展全额补贴项目

为了保障低收入群体享有的基本用电权益，解决居住在非洲偏远地区、急需解决用电需求、电价承受能力低的农村无电人口用电问题，推动非洲电力可及性，政府应该对离网项目前期建设和后期运营均进行一定的财政补贴。因此建议择优开展全额补贴项目，在项目规划时应重点惠及偏远村镇的居民用户，普惠性和示范意义强，满足低收入无电人口的最基本的用电需求。

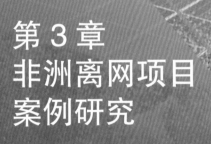

第 3 章
非洲离网项目
案例研究

本章以阿布贾农业高科技产业园区为研究对象，通过研究园区现状用电特性，结合"新能源+农业"的现代农业新理念，旨在为园区供电稳定、用能成本控制提出离网升级改造方案的概念性设计，并对园区离网升级改造方案的经济效益与社会效益进行研究，为非洲"新能源+农业"理念的推广提供参考方案与技术思路。

农业是非洲经济社会发展的重要支柱产业，是非洲大部分地区农村人口的主要收入来源。根据联合国粮食及农业组织（FAO）数据显示，撒哈拉以南非洲农业就业人口已占劳动力总量的 50% 以上，农业对非洲各国 GDP 的贡献率平均达 15%，而全球平均值仅为 4.1%。非洲大陆有大约 9.3 亿 hm^2 土地适合农业生产，约相当于美国国土面积。

然而非洲许多地区受制于经济条件以及缺乏可靠的电力供应，仍在使用传统的农耕设备，非洲农业发展的巨大潜力还远远没有得到释放。通过离网为非洲偏远农村提供生产用能，能够有效提高农业生产时间和效率，对于推动非洲经济社会发展具有重要的现实意义。为了让非洲离网项目更具代表性，本研究选取具有推广价值的离网农业项目——中地海外尼日利亚阿布贾农业高科技产业园区（以下简称"农业园区"）进行研究。

3.1 项目概况——阿布贾农业园区

农业园区基本概况

农业园区地理位置示意如图 3-1 所示，农业园区位于尼日利亚首都阿布贾中心商业区西北方向布瓦里地区乌沙法村，距离首都北外环高速公路约 11km，总占地面积 $89hm^2$。农业园区一面邻山丘，其余三面临公路，紧邻河流下游水库，具有较好的交通、水源、劳动力和社会治安条件。农业园区是西非地区首个现代化农业示范综合园区，主要工作包括种子种苗研发、农产品加工、园艺栽培、农资销售、农机培训和科研办公等。园区定位为以尼日利亚为中心辐射非洲的产业孵化园区；体现中国农业文化和文明的农业田园综合体以及中国在非洲的农业技术转移中心；促进非洲农业发展的知识信息学术交流和要素产品资源展览交易合作平台。

图 3-1　农业园区地理位置示意图

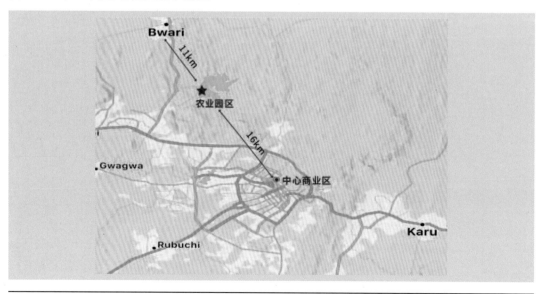

农业园区发展定位

如图 3-2 所示,农业园区发展定位包括农业展览交易合作平台、中非农业技术转移中心、现代化农业田园综合体等方面。

图 3-2　农业园区发展定位

农业园区功能区划

农业园区功能分区示意如图 3-3 所示,农业园区目前包括生活办公区和榨油厂等。其中,生活办公区位于农业园区的东北部,主要功能包括日常办公、生活服务等;榨油厂的主要功能是炼取食用油。农业园区未来还将投入使用一个技术示范中心,该中心包括能源动力中心、农产品加工中心、技术研发中心、技术培训中心等,主要功能包括园区发电供能、农产品加工、种子种苗研发、农机培训等。

图 3-3　农业园区功能分区示意图

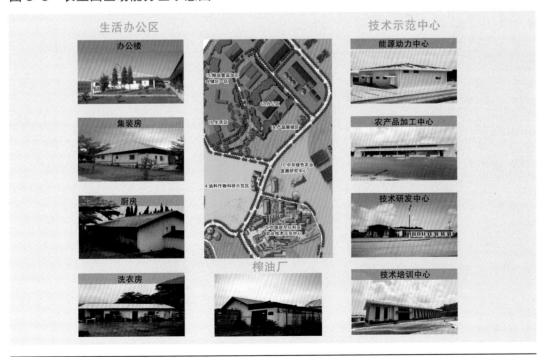

农业园区用电现状

农业园区当前年用电量约 323 MWh，用电设备功率合计约为 879.5 kW，日用电负荷峰值通常出现在白天，在 80～90 kW。农业园区用能示意如图 3-4 所示，农业园区主要包括生活办公区和榨油厂两个主要用电区。其中，生活办公区包括宿舍区和公共服务区，宿舍区的高峰期用电负荷约为 31 kW，公共服务区包括厨房、餐厅和洗衣房等，高峰期用电负荷约为 10 kW；榨油厂的高峰期用电负荷约为 45 kW。

目前，农业园区用电主要依靠市政供电，由 300 kVA 外电变压器接入。由于农业园区经常性停电，市政供电质量无法保证，为此农业园区采购了 3 台 150 kVA 的柴油发电机作为备用电源，3 台柴油发电机交替使用，每月柴油耗油量大约为 4000 L。随着柴油价格不断攀升，导致园区用电成本大幅上涨。鉴此，农业园区正计划引入新能源设备，建立自给自足的小型离网供电系统，为农业园区提供清洁电能，提高供电可靠性和经济性，并建设成为现代化的"新能源＋农业"示范项目。

图 3-4 农业园区用能示意图

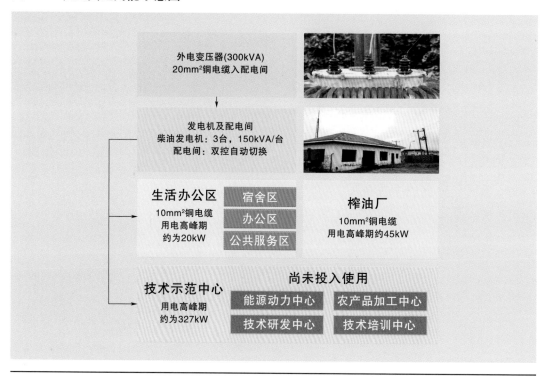

专栏："新能源＋农业"的应用

"新能源＋农业"是现代农业发展的重要方向之一。新能源具有清洁低碳、环境友好、经济性高、可再生且可就地获取并灵活布置的特点，特别适宜非洲农业人口分散性强的特点。采用"新能源＋农业"的现代化技术，不仅可以实现偏远地区各类农业生产场景下的能源稳定供应，还可以通过新能源设备因地制宜的灵活布置，实现农业生产的模式创新，推动新能源技术和农业生产的有效互补，有效提升偏远无电地区的农业生产效率，带动当地经济社会发展。

图 3-5　"新能源 + 农业"在非洲应用的 4 大优势

农业电力需求的增加与新能源发电成本的下降，促进了新能源在农业中的大规模运用。中国是全球农业大国和能源大国，随着中国"双碳"目标的提出，以能源体系清洁化、产业结构绿色化、消费方式电力化、生活方式低碳化为主要特征的经济社会绿色低碳转型将全面提速，"新能源 + 农业"模式也迎来发展的快车道，推动实现绿色、低碳、循环的农业高质量发展。

图 3-6　"新能源 + 农业"在中国常见的应用场景

光伏 + 农业种植

"光伏 + 农业种植"应用最为广泛，主要包括"光伏 + 设施园艺"和"光伏 + 大田种植"两种模式。其中，"光伏 + 设施园艺"将光伏发电与温室种植相结合，在满足作物生长光照的条件下，在棚顶安装太阳能光伏板，利用光伏发电，满足温室内的电能需求，充分利用大棚控光、控温和防病虫害的优势，生产出品质优良的产品；"光伏 + 大田种植"则是为了解决农业用地与产能用地之间矛盾的一种土地利用方式，实现一地多用，提高了土地的经济效益与利用效率。

案例：光伏发电领跑基地

中国陕西省铜川市国家光伏发电领跑基地装机规模 25 MW，占地面积约 6.33 km²，设计定位为"光伏发电 + 农业种植 + 观光旅游 + 扶贫"，除了满足周边地区用电需求和温室大棚供能，光伏面板下可种植药材。该电站不仅为当地生产生活提供了绿色电力，也推动了农业种植的新模式，在创造大量就业机会的基础上，促进当地群众脱贫增收，同时对黄土高原的生态修复发挥了积极作用。

光伏 + 畜禽养殖

"光伏 + 畜禽养殖"包括畜舍室内养殖与光伏露天养殖两种方式。其中，畜舍室内养殖利用光伏发电产生的电能调控畜舍内环境，如温度、湿度、排风装置等；光伏系统与畜禽的露天养殖结合形式是在光伏板下种植牧草，为畜禽提供充足养料的同时还为畜禽提供"荫凉"，避免过度暴晒，造成畜禽灼伤，这种光伏发电与畜禽养殖相结合的方式具有可观的能源效益，也是对农用土地资源的高效利用。

案例：光伏分布式牛棚电站

中国浙江省慈溪市牛棚分布式光伏电站位于浙江省慈溪市现代农业开发区，占地面积 0.8 km²，装机规模 15.6 MW。该项目在新建牛棚屋顶铺设光伏面板，棚顶光伏面板在夏季遮挡阳光的同时进行发电，为棚内畜禽创造舒适的内容环境，光伏发电产生的电能可用于舍内环境调控的电力来源。项目所发电量全部并入当地电网，为用户带来良好的经济效益与土地利用的综合效益。

光伏+提水灌溉

"光伏+提水灌溉"主要由太阳能电池阵列、逆变控制系统、提送水系统三部分组成。在旱季光照充足的条件下，转农业生产中不利条件为有利，运用光伏发电带动水泵、抽水机组等水利机械抽输水，实现抗旱灌溉，这种光伏运作模式已在许多有水无电地区实行，解决了这些地区的农田灌溉问题，保障了当地群众的切身利益。

案例：太阳能提水抗旱浇麦系统

中国河南省太阳能提水抗旱浇麦系统装机容量 7 kW，是中国首套太阳能有提水抗旱系统，光伏面板占地面积 77 m^2，输出功率 7680 W，提水扬程达 100 m 以上。项目在旱期运用太阳能产生的电能进行提水抗旱和农田灌溉，在非旱期将多余电量送入当地电网，以此获得项目的额外收益，该系统解决了旱期农田灌溉问题并提高了当地收入。

光伏+水产养殖

"光伏+水产养殖"是在水产养殖水域上方建设光伏发电系统，水上进行光伏发电，水下进行渔业养殖，这种模式被称为渔光互补型光伏电站。"渔光互补"模式既能充分利用空间、节约土地资源，又能利用光伏电站调节养殖环境，减少水体蒸发并抑制蓝藻繁殖，提高单位面积水域的经济价值与水体环境，具有良好的经济效益与环境效益。

案例：渔光互补光伏发电站

江苏省兴化市"渔光互补"光伏发电项目装机规模 178 MW，位于兴化市西北部沙沟镇。项目在鱼塘和湿地上方架设光伏组件进行发电，形成"上可发电、下可养鱼"模式，既充分利用空间、节约土地资源，又能利用光伏电站调节养殖环境，现已成为当地知名的多功能光伏农业产业基地，和一个光伏产业带动水产业及相邻产业综合发展的示范园区。

光伏 + 粮食加工

"光伏 + 粮食加工"最普遍的运用模式是在占地范围较大的粮仓仓顶铺设光伏面板。粮仓屋顶光伏电板具有保温隔热的效果，改善了粮仓内部储粮环境。通过粮仓屋顶光伏发电产生能源，为粮食储存系统中温控装置、湿控装置等运作装置提供动力，实现了粮仓低温储粮的可持续发展，节约了运作成本减少资源浪费，具有较好的经济效益。

案例：粮仓屋顶光伏发电站

漳州市草坂国家粮仓屋顶光伏发电站位于福建省漳州市西北部南靖县靖城镇，该项目利用草坂粮仓总面积 20000 m² 的标准化仓房屋顶，建设总装机容量 2 MW 的光伏发电组件。光伏发电产生的电能部分用于保持粮仓内温控系统、温控系统运作，其余电量可接入当地电网，项目充分利用全市粮库屋面优势资源，实现经济效益和科技储粮有效融合。

3.2 农业园区离网改造概念性设计

设计目标

解决园区供电问题	降低园区用能成本
以风电、光伏作为园区的供电主体，并通过配置电化学储能与柴油发电备用电源平滑新能源出力曲线，确保园区供电的可靠性，保证园区能够离网运行，解决园区的供电问题	通过建成"风，光，储，备"新能源发电系统，以低发电成本的新能源电代替柴油发电与市政供电，从而降低园区的用电成本
打造绿色、环保现代农业生产园区	在非洲推广现代化"新能源+农业"新技术
通过可再生能源的引入，促进园区的绿色、低碳、优质化建设，凸显新型农业技术中心的示范作用，打造绿色、环保的现代农业生产园区	推进本项目"新能源+农业"的新型应用模式，以典型案例的方式向全非洲进行辐射，在全非洲推广现代化"新能源+农业"新技术

设计原则

（1）既满足园区供电问题、降低用电成本，又体现了打造绿色农业园区的示范作用。

（2）运用近期方案解决现今用能需求，远期方案打造绿色"新能源+农业"应用典范。

（3）持续稳定型与环境友好型的能源供给，促进园区绿色环保与低碳排放建设。

（4）低发电成本的光伏新能源代替较高投入的发电形式，提高产业园区经济效益。

（5）以"新能源+农业"的应用典型方式辐射非洲，开启非洲农业创新经济模式的发展。

设计思路

通过对项目所在地可再生能源资源禀赋的调查，当地太阳能资源丰富，年总辐射量为 1889.8 kWh/m^2；风能资源较差，50 m 高度风速约 3.8 m/s，风功率密度约 60 W/m^2。经测算，采用光伏供电项目经济性最优。因此，农业园区离网升级改造以光伏作为供电电源，结合电化学储能，建设新能源离网供电系统。

图 3-7　光伏储能离网发电系统图

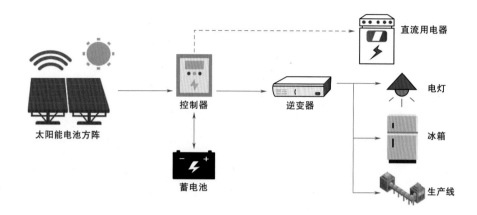

按照"总体规划，远近结合，效益优先"的思路，分别立足于农业园区生产生活现状和远期发展需要，离网供电升级改造按照"近期方案"和"远期方案"两种情景开展概念性设计。近期升级改造主要满足农业园区当前的生产生活用能需求，包括：生活办公区和榨油厂等，通过引入光伏发电，构建离网供电系统，解决农业园区当前缺电和经济性问题。远期升级改造主要着眼于技术示范中心投运后的生产生活用能需求，包括：生活办公区、榨油厂和技术示范中心等，通过构建光伏离网发电系统，并引入多元场景下的"新能源 + 农业"离网设备，满足农业园区用电需求，打造绿色、经济、低碳、环保、先进的现代化农业园区典范。离网供电改造近期和远期情景见表 3-1。

表 3-1　离网供电改造近期和远期情景

设计情景	设计定位	设计目标
近期方案	满足当前生产生活用能需求，包括：生活办公区和榨油厂等	解决农业园区当前缺电和经济性问题
远期方案	在满足当前生产生活用能需求的基础上，满足技术示范中心投运后的用能需求；同时，引入多元场景下的"新能源 + 农业"离网设备	打造绿色、经济、低碳、环保、先进的现代化农业园区典范

用电分析

对于近期方案，农业园区的用能分区主要包括生活办公区和榨油厂，其中生活办公区含有热水器 31 台、路灯 42 盏、空调 20 台、冰柜 / 冰箱 13 台、照明灯 183 盏、饮水机 8 台和消毒柜 4 台等；榨油厂含有榨油设备 1 套。上述设备总功率约为 136.72 kW，据农业园区工作人员现场实测，日均总用电量约为 911.2 kWh。其中，日间用电约为 684.4 kWh，夜间用电约为 226.8 kWh。近期情景设备用电统计见表 3-2。

表 3-2 近期情景设备用电统计

用能分区	设备	平均功率	数量	日均总用电量	日运行方式
生活办公区	热水器	1.5 kW	31 台	46.5 kWh	夜间运行 1 h
	路灯	100 W	42 盏	42 kWh	夜间运行 10 h
	空调	1.5 kW	20 台	360 kWh	平均 10 台全日运行
	冰柜/冰箱	200 W	13 台	62.4 kWh	全日运行
	照明灯	20 W	183 盏	18.3 kWh	夜间运行 5 h
	饮水机	420 W	8 台	6 kWh	随机运行 < 2 h
	消毒柜	800 W	4 台	16 kWh	随机运行 5 h
榨油厂	榨油设备	45 kW	1 套	360 kWh	日间运行 8 h
合 计		—	—	911.2 kWh	—

对于远期方案，农业园区的用能需求在近期方案的基础上，还需满足技术示范中心投运后的用能需求。该中心含有大米加工生产设备 8 台、辣椒粉生产设备 1 台、面包生产设备 8 台、培训教室用电设备 1 套和路灯 59 盏等。上述设备总功率约为 327 kW，远期方案设备总功率约为 463.72 kW，经测算，远期方案的日均总用电量约为 3816.48 kWh（近期方案 911.2 kWh + 远期新增 2905.28 kWh）。其中，日间用电量约为 3570.8 kWh，夜间用电量为 245.68 kWh（近期方案 226.8 kWh + 远期新增 18.88 kWh）。远期情景新增设备用电统计见表 3-3。

表 3-3 远期情景新增设备用电统计

用能分区	设备	平均功率	数量	日均总用电量	日运行方式
技术示范中心	大米加工生产	8.56 kW	8 台	684.8 kWh	日间运行 10 h
	辣椒粉生产	24 kW	1 台	240 kWh	日间运行 10 h
	面包生产	6.52 kW	8 台	521.6 kWh	日间运行 10 h
	培训教室用电	180 kW	1 套	1440 kWh	日间运行 8 h
	路灯	40 W	59 盏	18.88 kWh	夜间运行 8 h
合 计		—	—	2905.28 kWh	—

设计方案

♦ 近期方案的发电组合设计

离网光伏发电系统主要包括：光伏发电单元、储能单元和配电单元等。基于农业园区当前阶段的用电负荷运行特性，根据当地的太阳能资源情况，按照电力平衡和电量平衡的原则进行光伏和储能设计配比。配置方案见表3-4。

表3-4 光伏发电系统配置方案

序号	项目	技术规格	数量	单位	总价/万元
1	光伏发电单元	250 kW	1	套	100
2	储能单元	140 kW/280 kWh	1	套	46.2
3	配电单元	低压配电柜1面	1	套	5
4	国际运费	—	1	项	10
5	施工费用	—	1	项	10
6	项目管理费	—	1	项	10
合计		—	—	—	181.2

♦ 远期方案的发电组合设计

远期方案考虑到技术示范中心投运后，供冷、供热和供电需求显著增长，拟引入新型"冷热电三联供"技术。相比于传统新能源离网发电系统，新型"冷热电三联供"技术在新能源供电的同时，通过收集光伏余热向用户供热，并采用高能效比的制冷机组提高制冷效率，节约制冷用电。该技术可以有效提升项目的经济性，节约用电成本，提高能源综合利用效率。

专栏：新型"冷热电三联供"系统

"冷热电三联供"系统是指利用太阳能作为一次能源，在实现光伏发电的同时，重复利用产生的废热供热，并采用高能效比的制冷机组供冷的综合能源供应系统。该系统主要由光伏发电机组、电制冷装置、供热装置、控制装置四大模块构成，各模块协同作用，多能互补。在制冷环节，白天制冷机组运用光伏机组产生的电能进行制冷，将多余电量以冷量形式储存在蓄冷水箱中，夜间需要时可利用白天储存冷量为用户供冷，在这一过程中，制冷机组的COP能效比（制冷系数）可以达到5以上，达到高效制冷，降低制冷成本并提高经济效益。在供热环节，光伏热电组件在发电的同时，产生低温热能，将这一部分本应废置的热量进行充分利用，将其作为供热或生活热水的能量来源之一，同时以空气源热泵作为备用供热来源，保障不良天气状况下的供热稳定。该技术可以有效提高整个功能系统的一次能源利用率，实现了能源的梯级利用，有效提升了供能系统的经济性和综合效益。

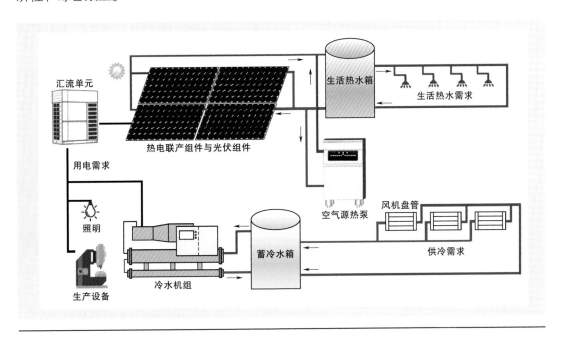

冷热电三联供系统主要包括：光伏发电单元、储能单元、制冷单元，制热单元和配电单元等。基于农业园区技术示范中心投入运营后的用电运行特性，并综合考虑用冷、用热需求，根据当地的太阳能资源禀赋，按照电力平衡、电量平衡和冷热供需平衡的原则进行冷热电三联供系统设计配比。配置方案见表3-5。

表3-5 冷热电三联供系统配置方案

序号	项目	技术规格	数量	单位	总价/万元
1	光伏发电单元	生活办公区 170 kW 技术示范中心 590 kW	1	套	380
2	储能单元	85 kW/170 kWh	1	套	36
3	制冷单元	生活办公区冷水机组 示范中心冷水机组 梯级蓄冷池	1	套	33.45
4	制热单元	生活热水箱 空气源热泵	1	套	0.85
5	配电单元	低压配电柜 1 面	1	套	5

续表

序号	项目	技术规格	数量	单位	总价/万元
6	国际运费	—	1	项	30
7	施工费用	—	1	项	80
8	项目管理费	—	1	项	30
合计		—	—	—	595.3

♦ 远期方案的离网产品设计

◆ 太阳能提水灌溉系统

远期方案拟引入 1 套太阳能提水灌溉系统，布置于蔬菜种植区，投资约为 25 万元。该系统利用太阳能拖动提水机自动浇灌作物，主要由光伏阵列、逆变器、潜水泵和蓄水箱或管网组成。相比于传统灌溉系统，该系统可以在没有电网接入的地方灵活布置，同时具有运行费用低、配套成本费用低、耗能排放少和提水效率高等综合优势。

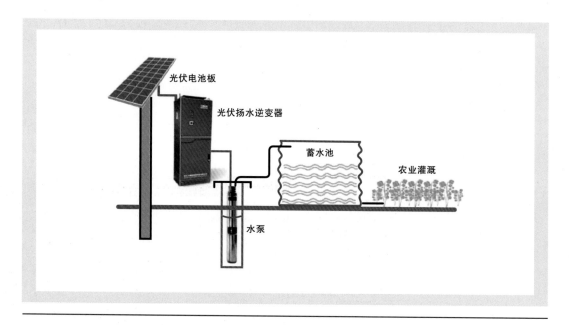

◆ 太阳能农业生产加工系统

远期方案拟引入 1 套太阳能农业生产加工系统，布置于农产品加工中心，投资约为 96 万元。该系统利用太阳能作为能源来源，给农业生产加工设备供电，主要由光伏阵列、逆变器、农业生产加工设备等组成，具有高效、清洁、低碳和经济的特点。

◆ 太阳能冷库

远期方案拟引入 1 套太阳能冷库，布置于实验室，投资约为 4 万元。太阳能冷库系统是利用太阳能进行制冷的系统。系统由光伏阵列、逆变器、集热器、冷凝器、蒸发器、储液器、阀门等组成，采用吸附式制冷系统，利用吸附床中的固体吸附剂对制冷剂周期性吸附，有效地实现了制冷循环，减少了能耗，提高了制冷效率，提高了经济性。

财务分析

项目计划由第三方投资，投资方与农业园区形成售购电关系，以此开展财务分析。由于项目投资总额较小，计划一次性投资，不考虑贷款和税收；项目无政府补贴，按10年运营期，储能电池更换0次进行测量分析运维费用约为总投资的0.1%（1～5年，后续每5年递增0.05%）；建设期为3个月；据调查，农业园区缴纳市政电费平均约为1.088元/kWh，以此作为基准测算电价，近期方案与远期方案的财务测算结果见表3-6和表3-7。

表3-6 近期方案财务测算表

序号	项目	数值
1	装机容量/MW	0.25
2	年发电量/MWh	367.03
3	总投资/万元	181.95
4	电费收入总额/万元	998.32
5	基准测算电价/（元/kWh）	1.088
6	项目投资回收期/年	6.70
7	项目投资财务内部收益率/%	15.32
8	总投资收益率（ROI）/%	9.04

表3-7 远期方案财务测算表

序号	项目	数值
1	装机容量/MW	0.76
2	年发电量/MWh	1153.65
3	总投资/万元	595.78
4	电费收入总额/万元	3034.9
5	基准测算电价/（元/kWh）	1.088
6	项目投资回收期/年	6.16
7	项目投资财务内部收益率/%	17.53
8	总投资收益率（ROI）/%	13.30

基于项目的投资、发电量和税率不变的情形下，项目的投资回收期和内部收益率在基准测算电价 -30%～0% 的变化区间内进行敏感性分析，数据见表 3-8。从数据分析可知，对于近期方案，随着基准测算电价在 -30%～0% 的区间内变化，投资回收期在 6～11 年，财务内部收益率在 7.42%～15.32%；对于远期方案，随着基准测算电价在 -30%～0% 的区间内变化，投资回收期在 6～9 年，财务内部收益率在 10.41%～17.53%。总体上看，采用可再生能源离网系统进行项目升级改造，近期方案和远期方案的电价较农业园区当前电价水平（即基准测算电价 1.088 元/kWh）有进一步下调空间，且同时满足项目投资回收期较短、内部收益率较高的要求，项目财务可行。

表 3-8 项目财务测算敏感性分析

基准测算电价变化幅度 /%	投资回收期 /a		项目投资财务内部收益率 /%	
	近期方案	远期方案	近期方案	远期方案
-30	10.13	8.82	7.42	10.41
-25	9.90	8.20	8.76	11.61
-20	8.97	7.67	10.09	12.81
-15	8.23	7.21	11.41	14.00
-10	7.63	6.81	12.72	15.81
-5	7.13	6.47	14.02	16.36
0	6.70	6.16	15.32	17.53

开发模式

通过对农业园区开展调研，农业园区用户为第一类用户，具有较大的用电需求以及较高的支付能力，用户侧综合可承受电价（$\overline{C}$）在 1.9 元/kWh 以上。经分析计算，引入新能源离网发电系统后，近期方案平准化度电成本（$LCOE$）约为 0.71 元/kWh，远期方案平准化度电成本（$LCOE$）约为 0.76 元/kWh。上述设计方案均满足 $\overline{C} \geqslant LCOE$，根据 2.3 关于开发模式的判定标准，若项目由第三方投资开发，均应采用盈利模式，电费收入可支持项目经济和可持续运营。

3.3 农业园区离网改造的综合效益

近期方案效益

效益类型	内　容	数量
经济效益	年电费预计减少	139104.4 元/a
环境效益	折算标准煤减少	45.2 t/a
环境效益	CO_2 排放量减少	144.6 t/a
社会效益	增加就业岗位	3 人
社会效益	推动可再生能源离网项目示范	—

远期方案效益

效益类型	内　容	数量
经济效益	年电费预计减少	359510.4 元/a
环境效益	折算标准煤减少	137 t/a
环境效益	CO_2 排放量减少	438.4 t/a
社会效益	增加就业岗位	7 人
社会效益	推动"新能源+农业"多元化应用	—

第 4 章
非洲离网项目推广探索

本章根据中国—非盟能源伙伴关系的指导思路，摸索可再生能源离网项目在非洲偏远地区的应用需求，探索"新能源＋农业"理念在非洲地区更多的实际应用场景，积极推动可再生能源离网项目在非洲其他领域的复制及推广，并对未来可再生能源离网项目产生的经济效益深入地分析与评估。促进解决非洲无电人口用电问题，实现能源高效利用，促进经济高质量发展。

根据前文分析，非洲地区约有 6.3 亿无电人口，大多分布在偏远的农村地区，可再生能源离网技术有望成为解决非洲无电人口用电问题的理想解决方案。为了通过可再生能源离网技术，促进联合国 2030 年人人享有可持续能源目标的实现，还需对非洲离网市场的项目资金需求进行分析，以完善和深化国际社会对于在非洲投资和推广离网技术可行性的认识。

通过对非洲地区的调查研究，并结合有关研究机构对于家庭用电的划分，设置低、中、高三种用电情景（表 4-1）。其中，低情景中，无电人口家庭通过可再生能源离网技术满足基本通电需求，即家庭拥有一些生活必需的小型负载设备，如电灯照明、收音机、电扇、电视机等，家庭户年均用电量在 50～300 kWh；中情景中，无电人口家庭通过可再生能源离网技术在满足基本通电需求的基础上，拥有少量电热和制冷负载设备，如电热壶、电磁炉、小型冰箱等，家庭户年均用电量在 300～1500 kWh；高情景中，无电人口家庭通过可再生能源离网技术能够较好满足现代社会日常生产和生活用电需求，拥有较多高功率负载设备，如电烤箱、烘干机、电热水器、空调等，家庭户年均用电量在 1500 kWh 以上。

表 4-1　用电情景分类

情景	定　　义	负荷参考/户	年均用电量/户
低	家庭拥有一些生活必需的小型负载设备，如电灯照明、收音机、电扇、电视机等	100～300 W	50～300 kWh
中	家庭在满足基本通电需求的基础上，拥有少量电热和制冷负载设备，如电热壶、电磁炉、小型冰箱等	300～1500 W	300～1500 kWh
高	家庭能够较好满足现代社会日常生产和生活用电需求，拥有较多高功率负载设备，如电烤箱、烘干机、电热水器、空调等	>1500 W	>1500 kWh

根据 IEA 数据，非洲地区家庭平均人口约为 6.9 人，6.3 亿无电人口约共有 0.9 亿户家庭，结合上述的情景分析与定义，采用可再生能源离网技术进行测算，三种场景下所需投资见表 4-2。采用可再生能源离网技术，在低情景下实现非洲电力可及所需总投资约为 1166.4 亿元，约合 164.3 亿美元；在中情景下实现非洲电力可及所需总投资约为 5832 亿元，约合 821.4 亿美元；在高情景下实现非洲电力可及所需总投资约为 14580 亿元，约合 2053.5 亿美元。

表 4-2　三种情景下可再生能源离网投资测算

情景	户均投资	总投资
低	1296 元（约 182.5 美元）	1166.4 亿元（约 164.3 亿美元）
中	6480 元（约 912.7 美元）	5832 亿元（约 821.4 亿美元）
高	16200 元（约 2281.7 美元）	14580 亿元（约 2053.5 亿美元）

第 5 章
结论与建议

5.1 结论

1. 撒哈拉以南非洲的电力可及是实现 SDG7 的攻坚任务

为了实现 SDG7 目标，全球无电人口数量从 2010 年的 12 亿下降到 2021 年的 7.33 亿。在此 10 年间，全球电力普及率显著上升，从 83% 上升到 91%。但世界各地区电力可及的推进并不均衡，世界上 75% 的无电人口生活在撒哈拉以南的非洲偏远地区，该地区无电人口基数大，且下降速度明显慢于世界其他地区，撒哈拉以南非洲的电力可及是实现 SDG7 的攻坚任务。当今人类社会是一个你中有我、我中有你的"命运共同体"已经成为普遍共识。在这样的时代背景下，撒哈拉以南非洲无电人口用电问题应该引起国际社会进一步关注和支持。

2. 新冠疫情蔓延对非洲实现电力可及带来了严峻挑战

受新冠疫情影响，全球经济发展低迷，特别是在电气化方面，已经取得的良好进展开始放缓甚至退步，对撒哈拉以南非洲影响更为明显。2020 年该地区无电人口数量在连续 5 年下降以后首次呈现增长，2021 年该地区无电人口数量进一步上升到 5.97 亿，较 2019 年增幅达 4%。新冠疫情的反复跌宕不仅减缓了非洲无电地区主干电网的接入速度，也使许多非洲国家原本制定的离网接入计划明显滞后，新冠疫情的暴发和蔓延导致生产生活供应链中断，投资和建设放缓，经济收入下降，给后疫情时代的非洲电力可及之路带来了严峻挑战。

3. 离网是解决非洲无电人口用电实际困难的理想方案

2021 年撒哈拉以南非洲地区有约 5.9 亿无电人口，这部分人大多数生活在非常贫困的农村，地处偏远，交通阻塞，基础设施薄弱，给电网施工建设带来了很大不便。同时，非洲地区电网较为薄弱，覆盖率低，导致主网延伸的总投资较大，难以在市场商业原则下有效解决非洲无电地区人口的用电问题。非洲可再生能源资源丰富，可循环利用并就地获取，采用可再生能源供电的离网系统不仅在规模上和位置上可以布置灵活，而且技术成熟、总投资小，符合非洲无电地区的实际情况，是当前经济技术条件下解决无电人口用电问题的理想方案。

4. 离网成功的关键是根据实际情况选择合适的开发模式

良好的经济性是电力项目可持续开发和运营的关键因素。非洲离网项目的受众多分布在偏远地区，经济基础薄弱，项目所在地的用户侧电价承受能力通常较弱，仅依靠当地居民电费收入通常难以支持离网项目的经济性。这就需要在前期对项目做好评估，根据离网项目覆盖区域的发电条件、用电需求、各类用户的电价承受能力综合评估是否需要外部补贴或捐助，以及所需的金额，从而据实选择合适的开发模式，包括盈利模式、部分补贴模式和全额补贴模式，确保项目具备可持续运行条件，同时保障低收入群体享有普惠性的较低电价。

5．"新能源＋农业"离网供电在非洲具有广阔的应用前景

农业是非洲大陆的经济支柱产业，也是非洲许多地区的人口主要收入来源。采用"新能源＋农业"模式，一方面可以利用新能源灵活布置的特点，实现非洲不同地区各类农业生产场景的稳定供电；另一方面可以根据农业生产的特点，与新能源技术相结合，通过农业经营设施合理嫁接新能源，发挥互补效应，实现经营模式创新，实现一地多用，提高单位土地产出率，从而有效提升非洲大陆，特别是偏远地区，农业生产效率，带动非洲当地经济社会发展。非洲耕地辽阔，"新能源＋农业"可有效释放农业生产潜力，具有较大的应用推广前景。

5.2 建议

非洲无电人口众多，涉及因素复杂交织。新冠肺炎疫情全球大流行使无电人口问题更加严峻，也使得 2030 年联合国可持续发展目标的实现增添了风险和不确定性，非洲电力可及之路依然任重道远。面对挑战，人类社会比任何时候都更需要加强合作，共克时艰，携手前行。国际社会，包括国际组织、政府部门、行业协会、金融机构、能源企业等利益攸关方，必须团结起来，从资金、市场、技术、能力、创新和推广等各个方面做好携手应对挑战的准备，为非洲无电地区人民追求幸福生活、人类社会可持续发展作出更加积极的努力。

1. 加强对非电力可及投融资支持

非洲和国际社会当务之急是为所有非洲人提供负担得起的现代能源，但目前仍然缺乏资金保障，根据 IEA 数据，非洲若实现 SDG7 需每年投资 250 亿美元。因此需要依托国际组织、金融机构、公共和私营部门加强对非电力可及投融资支持，增加国际公共资金流动并以更公平的方式分配资金，以实现不让任何人掉队的全球电力可及的未来。

2. 做好离网市场调研和项目论证

离网项目的成功需要做好离网市场调研和项目论证。为了实现离网项目的经济可持续运营，从市场整体上看，需要因地制宜，统筹开发，有序、逐步地进行离网市场调研。从项目层面上看，经济性是离网项目成功的关键，需要选择合适的开发模式，做好项目论证。建议大力推进盈利项目，适当支持部分补贴项目，择优开展全额补贴项目。

3. 提升电力技术和管理能力建设

目前非洲在可再生能源领域仍存在技术人才缺乏、项目建设和管理能力不足等问题。为推动非洲电力可及工作，需要积极提升非洲电力技术，开展技术研讨，促进国际社会与非洲的务实合作，促进本土可再生能源技术发展。还需积极落实非洲可再生能源能力建设培训，多形式推动非洲人才培养工作，帮助非洲提升可再生能源项目建设和管理水平。

4. 推广新技术提升经济社会效益

非洲的可再生能源资源禀赋优异，需要因地制宜地开发并结合创新性的技术方案。不断提高发电效率，降低发电成本，促进项目的经济性，发挥更大的社会效益。如冷热电三联供设备，减少蓄电池使用的同时可以实现供冷供热供电，降低项目成本，增加绿色发电，减少碳排放量。促进可再生能源行业多样化发展，为非洲带来更多的就业机会。

5. 推动"小而美"的示范项目开发

非洲发展可再生能源潜力巨大，应当充分利用其能源优势，通过可再生能源与农业、采矿、冶金等产业结合，推动可再生能源应用场景的创新。在满足非洲当地亟须的前提下，大力开发投入小、建设快、惠民生的可再生能源离网项目，发挥好这类"小而美"的项目的示范引领作用，以点带面，实现在非洲的逐步推广，助力非洲实现电力可及目标。

声 明

受中国国家能源局委托，水电水利规划设计总院（以下简称"水电总院"）牵头配合开展中国—非盟能源伙伴关系的筹建和运行。推动非洲电力可及是中国—非盟能源伙伴关系的重要任务，也是本研究的重点内容。本研究对于非洲电力可及的分析预测仅建立在离网技术领域当前阶段所掌握的资料之基础上，对于因政策、安全和环境等外部因素导致的无法预见情况，建议根据实际情况酌情加以判断。

本研究的知识产权由水电总院及世界资源研究所共同拥有，未经许可，任何单位或个人不得以任何形式复制转载，不得将本研究进行产品传播或用于任何商业、营利目的。本研究所采用的数据全部来自权威机构发布（或正规出版社出版）的书籍、期刊、杂志等，并对不同机构发布的同类数据进行了对比，尽可能确保所采用数据的准确性。因原机构发布数据的误差或错误导致本研究结果出现偏差，水电总院不承担相关责任。

注：本研究所涉及的图片除个别未找到出处的图片外，均已购买图片授权，享受图片的使用权。若后续未标明出处的图片涉及版权问题，请及时与水电水利规划设计总院取得联系。

STATEMENT

Entrusted by the National Energy Administration, the China Renewable Energy Engineering Institute (hereinafter referred to as "CREEI") took the lead in cooperating in the preparation and operation of China-AU Energy Partnership. As an important task of AU-China Energy Partnership, promoting electricity access in Africa is also the key content of this study. The analysis and forecast of electricity access in Africa in this study is only based on the information available in the field of off-grid technology at the present phase. For unforeseen situations caused by external factors such as policy, security and environment, it is suggested to make judgement according to the actual situation.

The intellectual property rights for this study are jointly owned by CREEI and the World Resources Institute (WRI China). Without permission, no unit or individual may copy and reprint this study in any form, or disseminate this study or use it for any commercial or profit-making purposes. All the data used in this study come from books, periodicals and magazines published by authoritative organizations (or regular publishing houses), and the similar data published by different organizations are compared to ensure the accuracy of the data used as much as possible. The CREEI does not bear relevant responsibilities for the results of this study deviated due to the error of the data released by the original organization.

Note: All the pictures involved in this study have been authorized to use and enjoy the right to use the pictures, except for some pictures whose source is found. If the subsequent image without the source is involved in copyright issues, please contact our institute in time.

5.Promote the development of "small and beautiful" demonstration project

There is great potential in Africa to develop renewable energy, thus it shall make full use of its energy advantages and promote the innovation of renewable energy application scenarios by combining renewable energy with agriculture, mining, metallurgy and other industries. On the premise of meeting the local urgent needs in Africa, it is necessary to vigorously develop renewable energy off-grid projects with small investment, fast construction and benefits to people's livelihood, and give full play to the demonstration and leading role of such "small and beautiful" projects, so as to realize the gradual promotion in Africa by fanning out from point to area and help Africa achieve the goal of electricity access.

2.Carry out off-grid market research and project demonstration

The success of off-grid projects requires off-grid market research and project demonstration. To realize the economic sustainable operation of off-grid projects, from the perspective of the market as a whole, it is necessary to adjust measures to local conditions, make overall plans for development, and conduct off-grid market research in an orderly and gradual manner. At the project level, economy is the key to the success of off-grid projects, thus it is necessary to choose a suitable business model and conduct project demonstration. It is suggested to vigorously promote profitable projects, appropriately support some subsidized projects, and carry out fully subsidized projects on the basis of merit.

3.Improve the power technology and management capacity building

At present, there are still some problems about renewable energy in Africa, such as lack of technical personnel, insufficient project construction and management capacity. To promote electricity access in Africa, it is necessary to actively upgrade power technology in Africa, conduct technical discussions, and promote practical cooperation between the international community and Africa and the technological development of local renewable energy. It is also necessary to actively implement the training of renewable energy capacity building in Africa, promote the training of African talents in various forms, and help Africa improve the construction and management level of renewable energy projects.

4.Popularize new technologies and improve economic and social benefits

Africa is endowed with excellent renewable energy resources, which need to be developed according to local conditions and combined with innovative technical solutions. It is necessary to continuously improve power generation efficiency, reduce power generation costs, promote the economy of the project, and bring greater social benefits into play. For example, combined cooling, heating and power equipment can realize cooling, heating and power supply while reducing the use of batteries, reduce project costs, increase green power generation and reduce carbon emissions. It can promote the diversified development of the renewable energy industry and bring more employment opportunities to Africa.

production scenarios in different regions of Africa; on the other hand, according to the characteristics of agricultural production, and combined with new energy technology, agricultural management facilities are rationally grafted with new energy to exert complementary effects, realize business model innovation and multi-purpose in one place, and enhance the output rate per unit land, thus effectively improving agricultural production efficiency on the African continent, especially in remote areas, and driving local economic and social development in Africa. Africa enjoys vast cultivated land, and "new energy + agriculture" can effectively release agricultural production potential, with great application and popularization prospects.

5.2 SUGGESTIONS

There is a large population without access to electricity in Africa, and the factors involved are complex and intertwined. The global pandemic of COVID-19 has made the problem of people without access to electricity more serious, and also added risks and uncertainties to the realization of the United Nations Sustainable Development Goals by 2030. There is still a long way to go to reach electricity access in Africa. Facing challenges, human society needs to strengthen cooperation more than ever, overcome difficulties and move forward hand in hand. The international community, including international organizations, government departments, industry associations, financial institutions, energy enterprises, and other stakeholders, must band together and be prepared to meet challenges hand in hand from capital, market, technology, capability, innovation and promotion, so as to make more active efforts for people in African regions without access to electricity to pursue a happy life and sustainable development of human society.

1.Strengthen investment and financing support for electricity access in Africa

The urgent task for Africa and the international community is to provide affordable modern energy for all Africans, but there is still a lack of financial guarantee. According to IEA data, it needs to invest USD 25 billion every year if Africa realizes SDG7. Therefore, it is necessary to rely on international organizations, financial institutions, public and private sectors to strengthen investment and financing support for non-electricity access, increase international public capital flows and allocate funds in a more equitable way, so as to realize the future of global electricity access that no one is left behind.

in 2021. Most of them live in very poor rural areas, with remote location, traffic block and weak infrastructure, which brings great inconvenience to power grid construction. Meanwhile, the power grid in Africa is weak and the coverage rate is low, which leads to a large total investment in the extension of the main power grid, and it is difficult to effectively solve the power consumption problem of the population without access to electricity in Africa under the principle of market commerce. Africa is rich in renewable energy resources, which can be recycled and obtained locally. The off-grid system powered by renewable energy is not only flexibly arranged in scale and location, but also mature in technology and small in total investment, which is in line with the actual situation of African regions without access to electricity. It is an ideal scheme to solve the power consumption problem of population without access to electricity under the current economic and technological conditions.

4. The key to the success of off-grid is to choose an appropriate business model in light of the actual situation

Favorable economy is the key factor of sustainable development and operation of power projects. The recipients of off-grid projects in Africa are distributed in remote areas, with a weak economic foundation. The electricity price bearing capacity at the user side where the project is located is usually weak, and it is usually difficult to support the economy of off-grid projects only by relying on the electricity revenue of local residents. Therefore, it is necessary to evaluate the project in the early stage. In view of the power generation conditions, power demand and electricity price bearing capacity of various users in the off-grid project coverage area, whether external subsidies or donations and amount are needed is comprehensively assessed, so as to choose the appropriate business model in light of the facts, including the profit model, the partial subsidy model and the full subsidy model, to ensure sustainable operation conditions for the project, and guarantee that low-income groups enjoy inclusive lower electricity prices at the same time.

5. "New energy + agricultural" off-grid power supply has broad application prospects in Africa

Agriculture serves as the pillar industry of the African continent and the main source of incomes for the population in many parts of Africa. In terms of adopting the "new energy + agriculture" model, on the one hand, it can take advantage of the flexible arrangement of new energy to achieve stable power supply for various agricultural

5.1 CONCLUSIONS

5.1 CONCLUSIONS

1.Electricity access in Sub-Saharan Africa is a key task to achieve SDG7

To achieve the SDG7, the number of people without access to electricity in the world dropped from 1.2 billion in 2010 to 730 million in 2021. During the 10 years, the global power penetration rate has increased significantly from 83% to 91%. However, the promotion of electricity access in different regions of the world is uneven. About 75% of the global population without access to electricity lives in remote areas of Sub-Saharan Africa, where there is large population base without access to electricity and its decline rate is obviously slower than that in other parts of the world. Electricity access in Sub-Saharan Africa is a crucial task to realize the SDG7. It has become a common consensus that today's human society is a "community of destiny" with a bit of me in you and a bit of you in me. Against this background, the problem of electricity consumption of people without access to electricity in Sub-Saharan Africa shall be further concerned and supported by the international community.

2.The spread of COVID-19 has brought severe challenges to electricity access in Africa

Influenced by COVID-19, the global economic development is sluggish, especially in electrification, and the good progress made has begun to slow down or even regress, with more apparent impact on Sub-Saharan Africa. The number of people without access to electricity in this region increased for the first time in 2020 after declining for five consecutive years, while that of people without access to electricity in this region further increased to 597 million in 2021, an increase of 4% compared with that in 2019. The repeated ups and downs of COVID-19 not only mitigate the access speed of the backbone power grid in African regions without access to electricity, but also make the off-grid access plan originally formulated by many African countries lag behind obviously. The outbreak and spread of COVID-19 have cut down production and living supply chains, hindered investment and construction, and reduced economic incomes, bringing severe challenges to electricity access in Africa in the post-pandemic era.

3.Off-grid is an ideal solution to solve practical difficulties of African population without access to electricity

There were about 590 million people without access to electricity in Sub-Saharan Africa

5

**CONCLUSIONS
AND SUGGESTIONS**

Table 4-2 Calculation of investment in renewable energy off-grid projects under three scenarios

Scenarios	Average investment per household	Total investment
Low	RMB 1296 (about USD 182.5)	RMB 116.64 billion (about USD 16.43 billion)
Medium	RMB 6480 (about USD 912.7)	RMB 583.2 billion (about USD 82.14 billion)
High	RMB 16200 (about USD 2,281.7)	RMB 1.458 trillion (about USD 205.35 billion)

Continued

Scenario	Definition	Load reference/ household	Average annual electricity consumption/household
Medium	On the basis of meeting the basic power demand, families have a small number of electric heating and refrigeration load equipment, such as electric kettles, induction cookers, small refrigerators and etc	300 ~ 1500 W	300 ~ 1500 kWh
High	Families can better meet the daily production and living electricity demand of modern society, and possess more high-power load equipment, such as electric ovens, dryers, electric water heaters, air conditioners, etc	> 1500 W	> 1500 kWh

According to IEA data, there are about 6.9 persons per African household on average, and there are 630 million people in about 90 million households without access to electricity. Combined with the above scenario analysis and definition, the renewable energy off-grid technology is used for calculation, with the investment required under the three scenarios shown in Table 4-2. Using renewable energy off-grid technology, the total investment required to achieve electricity access in Africa under low scenario is about RMB 116.64 billion, or about USD 16.43 billion; the total investment required to achieve electricity access in Africa under the medium scenario is about RMB 583.2 billion, or about USD 82.14 billion; the total investment required to achieve electricity access in Africa under the high scenario is about RMB 1.458 trillion, or about USD 205.35 billion.

According to the analysis in previous chapters, there are about 630 million people without access to electricity in Africa, most of which are distributed in remote rural areas. Renewable energy off-grid technology is expected to become an ideal solution of the power consumption problem of people without access to electricity in Africa. To promote the realization of the United Nations goal of Sustainable Energy For All in 2030 through the off-grid technology of renewable energy, it is also necessary to analyze the project funding needs of the off-grid market in Africa, so as to improve and deepen the international community's understanding of the feasibility of investing and promoting off-grid technology in Africa.

Through the investigation and study in Africa, combined with the division of household electricity consumption by relevant research institutions, three scenarios of low, medium and high electricity consumption are set, as shown in Table 4-1. In the low scenario, families without access to electricity meet the basic power demand through renewable energy off-grid technology, that is, families possess some small load equipment necessary for life, such as electric lighting, radio, electric fan, TV set, etc., and the average annual electricity consumption of household is 50~300kWh; in the middle scenario, households without access to electricity possess a small number of electric heating and refrigeration load equipment, such as electric kettles, induction cookers, small refrigerators, etc., on the basis of meeting the basic power demand through renewable energy off-grid technology, and the average annual electricity consumption of households is 300~1500kWh; In the high scenario, households without access to electricity can better meet the daily production and living electricity demand of modern society through renewable energy off-grid technology, and possess more high-power load equipment, such as electric ovens, dryers, electric water heaters, air conditioners, etc. The average annual electricity consumption of household is over 1500kWh.

Table 4-1　Classification of electricity consumption scenario

Scenario	Definition	Load reference/ household	Average annual electricity consumption/household
Low	Families have some small load equipment necessary for life, such as electric lighting, radio, electric fan, TV set and etc	100 ~ 300 W	50 ~ 300 kWh

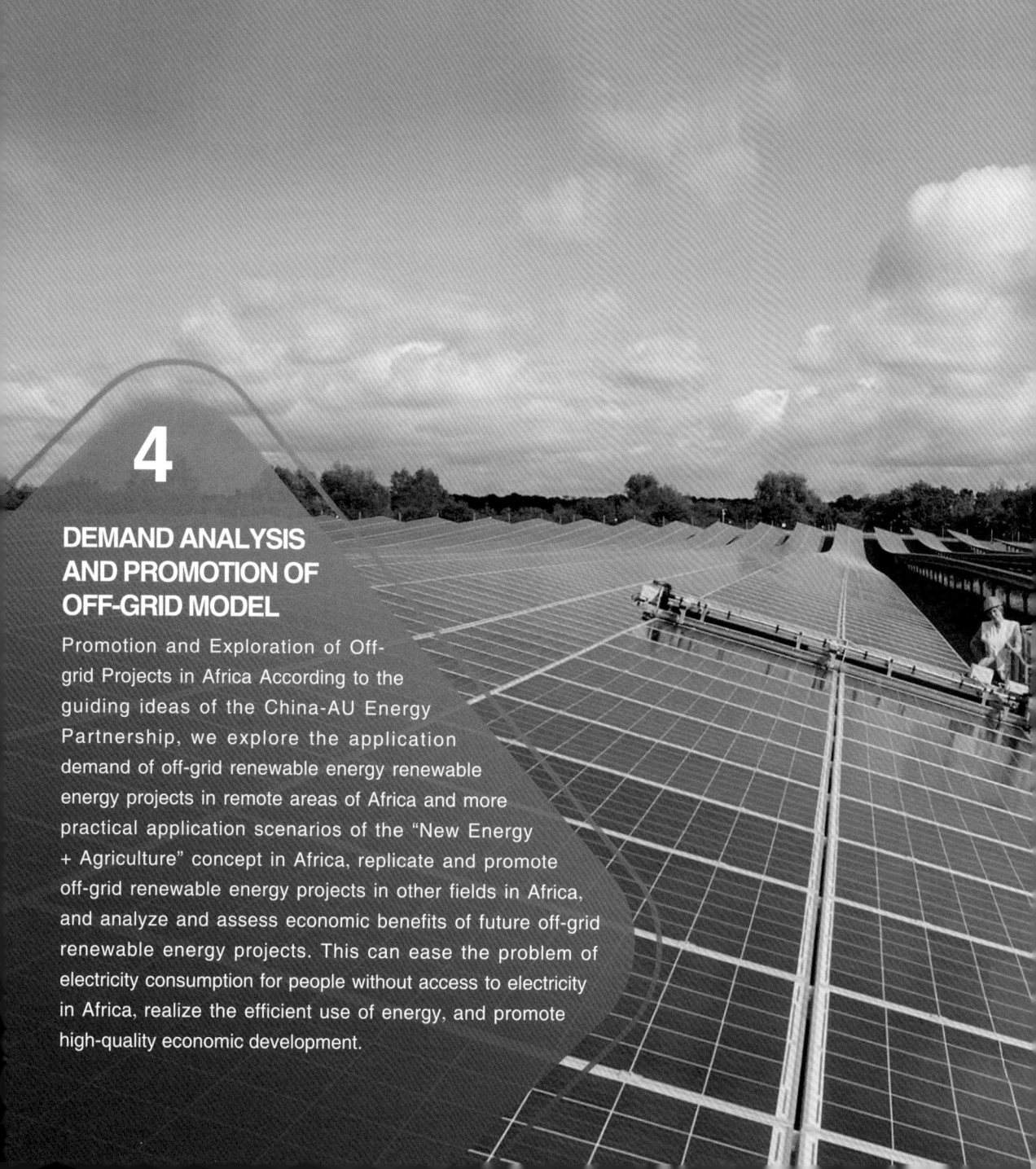

4

DEMAND ANALYSIS AND PROMOTION OF OFF-GRID MODEL

Promotion and Exploration of Off-grid Projects in Africa According to the guiding ideas of the China-AU Energy Partnership, we explore the application demand of off-grid renewable energy renewable energy projects in remote areas of Africa and more practical application scenarios of the "New Energy + Agriculture" concept in Africa, replicate and promote off-grid renewable energy projects in other fields in Africa, and analyze and assess economic benefits of future off-grid renewable energy projects. This can ease the problem of electricity consumption for people without access to electricity in Africa, realize the efficient use of energy, and promote high-quality economic development.

Continued

Benefit type	Content	Quantity
(8) Social benefits	increase of employment posts	3 persons
	Promotion of the demonstration of renewable energy off-grid project	—

Benefits of long–term scheme

Benefit type	Content	Quantity
(¥) Economic benefits	Expected decrease of annual electric charge	RMB 359510.4/a
◯ Environmental benefits	Reduction of converted standard coal	137t/a
	CO_2 emission reduction	438.4t/a
(8) Social benefits	Increase of employment posts	7 persons
	Promotion of the diversified application of "new energy +agriculture"	—

Continued

Benchmark electricity price change range /%	Pay-back period /a		Financial internal rate of return of project investment /%	
	Near-term scheme	Long-term scheme	Near-term scheme	Long-term scheme
-20	8.97	7.67	10.09	12.81
-15	8.23	7.21	11.41	14.00
-10	7.63	6.81	12.72	15.81
-5	7.13	6.47	14.02	16.36
0	6.70	6.16	15.32	17.53

Business model

Through the investigation of agricultural park, the users of agricultural park are the first class users with great demand for electricity and high ability to pay, and the comprehensive affordable electricity price $(\overline{C})$ on the user side is above RMB 1.9/kWh. After analysis and calculation and the introduction of new energy off-grid power generation system, the *LCOE* of the near-term scheme is about RMB 0.71/kWh, while that of the long-term scheme is about RMB 0.76/kWh. All the above design schemes meet $\overline{C} \geqslant LCOE$. According to the judgment standard of business model in Chapter 2.3, the profit model shall be adopted if the project is invested and developed by a third party, and the electric charge income can support the economical and sustainable operation of the project.

3.3 COMPREHENSIVE BENEFITS OF OFF-GRID TRANSFORMATION OF THE AGRICULTURAL PARK

Benefits near–term scheme

Benefit type	Content	Quantity
(¥) Economic benefits	Expected decrease of annual electric charge	RMB 139104 .4/a
⬭ Environmental benefits	Reduction of converted standard coal	45.2t/a
	CO_2 emission reduction	144.6t/a

Continued

S/N	Item	Unit	Value
3	Total investment	RMB 10,000	595.78
4	Total electric charge income	RMB 10,000	3034.9
5	Benchmark electricity price	RMB/kWh	1.088
6	Pay-back period of project	Year	6.16
7	Financial internal rate of return of project investment	%	17.53
8	Rate of return (ROI) of total investment	%	13.30

Based on the constant investment, power generation and tax rate of the project, the sensitivity analysis of the pay-back period and internal rate of return of the project is carried out within the change range of -30% ~ 0% of the benchmark electricity price, with the data shown in Table 3-8. From the data analysis, it can be seen that for the near-term scheme, with the change of benchmark electricity price in the range of -30% ~ 0%, the pay-back period of investment is 6 ~ 12 years, and the financial internal rate of return is 7.42% ~ 15.32%; For the long-term scheme, with the change of benchmark electricity price in the range of -30% ~ 0%, the pay-back period of investment is 6 ~ 9 years, and the financial internal rate of return is 10.41% ~ 17.53%. Generally speaking, the renewable energy off-grid system is adopted to upgrade the project. Compared with the current electricity price level of the agricultural park (i.e., the benchmark electricity price is RMB 1.088/kWh), the electricity price of the near-term scheme and the long-term scheme has room for further reduction, and meets the requirements of short pay-back period and high internal rate of return of the project at the same time, thus the project is financially feasible.

Table 3-8　Sensitivity analysis of project financial calculation

Benchmark electricity price change range /%	Pay-back period /a		Financial internal rate of return of project investment /%	
	Near-term scheme	Long-term scheme	Near-term scheme	Long-term scheme
-30	10.13	8.82	7.42	10.41
-25	9.90	8.20	8.76	11.61

Financial analysis

The project plans to be invested by a third party, and the investor and the agricultural park form a relationship between electricity seller and buyer, so as to carry out financial analysis. It is planned to make a one-time investment without considering loans due to low investment on the project. The project has no government subsidy. According to the 10-year operation period, the operation and maintenance cost of energy storage battery replacement is about 0.1% of the total investment (1~5 years, increasing by 0.05% every 5 years). The construction period is 3 months; According to the survey, the average local municipal electricity price is about 1.088 RMB /kWh, which is used as the benchmark to calculate the electricity price. Financial analysis and calculation are carried out under this boundary condition, and the financial calculation results of near-term scheme and long-term scheme are shown in Table 3-6 and Table 3-7 respectively.

Table 3-6   Financial calculation for near-term scheme

S/N	Item	Unit	Value
1	Installed capacity	MW	0.25
2	Annual power generation	MWh	367.03
3	Total investment	RMB 10,000	181.95
4	Total electric charge income	RMB 10,000	998.32
5	Benchmark electricity price	RMB/kWh	1.088
6	Pay-back period of project	Year	6.70
7	Financial internal rate of return of project investment	%	15.32
8	Rate of return (ROI) of total investment	%	9.04

Table 3-7   Financial calculation for long-term scheme

S/N	Item	Unit	Value
1	Installed capacity	MW	0.76
2	Annual power generation	MWh	1153.65

energy as energy source to supply power to agricultural production and processing equipment, which is mainly composed of PV array, inverter and agricultural production and processing equipment, and is characterized by high efficiency, cleanliness, low carbon and economy.

◆ Solar energy cold storage

In the long-term scheme, it is planned to introduce 1 set of solar energy cold storage, which will be arranged in the laboratory with an investment of about RMB 40000. Solar energy cold storage system is a system that uses solar energy for refrigeration. The system is composed of PV array, inverter, collector, condenser, evaporator, liquid reservoir, valve, etc. Adsorption refrigeration system is adopted, and the solid adsorbent in the adsorption bed adsorbs the refrigerant periodically, which effectively realizes the refrigeration cycle, reduces energy consumption, and improves refrigeration efficiency and economy.

Continued

S/N	Item	Technical specifications	Quantity	Unit	Total price/ RMB10,000
6	International freight	—	1	Item	30
7	Construction cost	—	1	Item	80
8	Project management fee	—	1	Item	30
	Total	—	—	—	595.3

◆ Off–grid product design of long–term scheme

◆ Solar energy irrigation by processing system

In the long-term scheme, it is planned to introduce 1 set of solar energy irrigation by pumping system, which will be arranged in the vegetable planting area with an investment of about RMB 250000. The system uses solar energy to drive water pump to irrigate crops automatically, which is mainly composed of PV array, inverter, submersible pump and water storage tank or pipe network. Compared with the traditional irrigation system, this system can be flexibly arranged in places without power grid access, and has the comprehensive advantages of low operation cost, low supporting cost, less energy consumption and emission and high water pumping efficiency.

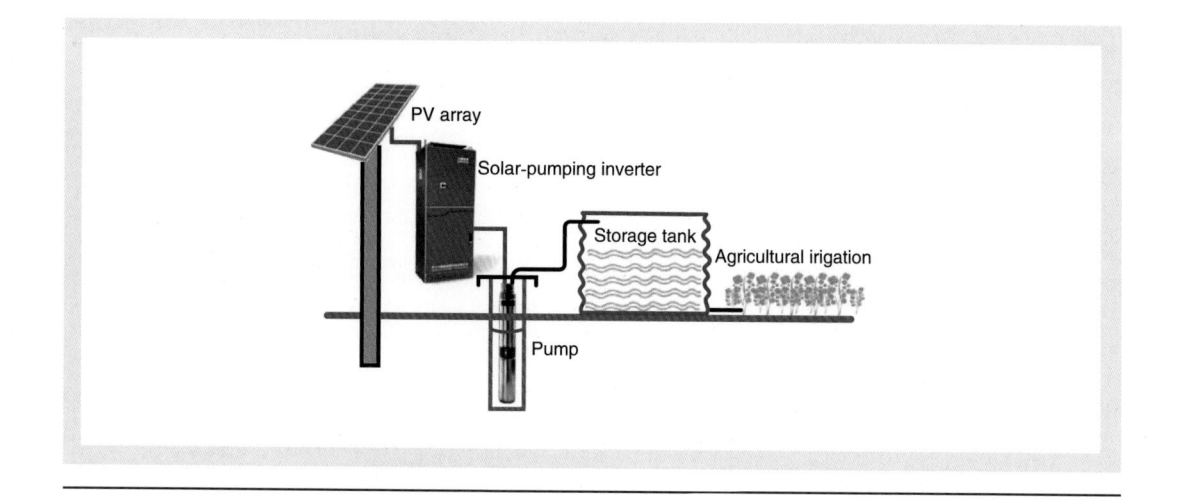

◆ Solar energy agricultural production and processing system

In the long-term scheme, it is planned to introduce 1 set of solar energy agricultural production and processing equipment, which will be arranged in the agricultural product processing center with an investment of about RMB 960000. This system uses solar

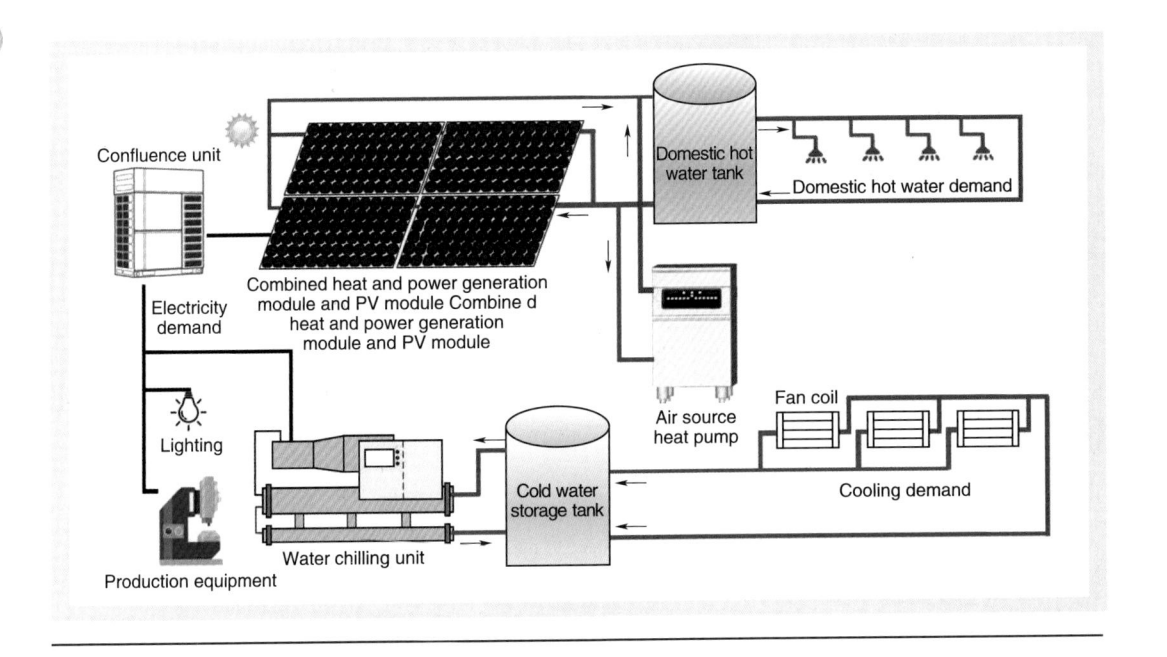

Combined Cooling, Heating and Power system mainly includes PV power generation unit, cooling unit, heating unit, energy storage unit and distribution unit. Based on the power operation characteristics after the agricultural park technology demonstration center is put into operation, and considering the demand of cooling and heat, the design ratio of combined cooling, heating and power system is carried out according to the local solar energy resource endowment and the principle of power balance, electricity balance and cooling and heat supply and demand balance. The configuration scheme is shown in Table 3-5 below.

Table 3-5   Configuration of combined cooling, heating and power system

S/N	Item	Technical specifications	Quantity	Unit	Total price/ RMB10,000
1	power generation unit	170kW in living and office area 590kW in technology demonstration center	1	Set	380
2	Energy storage unit	85 kW/170 kWh	1	Set	36
3	Cooling unit	Water chilling unit in living and office are Water chilling unit in demonstration center Cascade cold storage pool	1	Set	33.45
4	Heating unit	Domestic hot water tank Air source heat pump	1	Set	0.85
5	Distribution unit	1 low voltage distribution cabinet	1	Set	5

◆ Power generation combination design of long-term scheme

In the long-term scheme, considering that the demand for cooling, heating and power supply has increased significantly after the technology demonstration center is put into operation, it is planned to introduce a new "Combined Cooling, Heating and Power" technology. Compared with the traditional new energy off-grid power generation system, the new "Combined Cooling, Heating and Power" technology provides heat for users by collecting PV waste heat while supplying power with new energy, and refrigeration units with high energy efficiency ratio is adopted to improve refrigeration efficiency and save refrigeration electricity. This technology can effectively improve the economy of the project, save electricity cost and improve the comprehensive utilization efficiency of energy.

COLUMN: NEW "COMBINED COOLING, HEATING AND POWER" SYSTEM

"Combined Cooling, Heating and Power" system is a comprehensive energy supply system that uses solar energy as primary energy, reuses waste heat for heating while realizing PV power generation, and adopts refrigeration units with high energy efficiency ratio for cooling. The system is mainly composed of four modules: PV generator unit, electrical refrigeration device, heating device and control device. Each module works synergistically, with multi-energy complement. In the refrigeration link, the refrigerating unit uses the electric energy generated by PV units for refrigeration during the day. The surplus electricity is stored in the cold water storage tank in the form of cold energy, and the cold energy stored during the day can be used for cooling for users when needed at night. In this process, the COP energy efficiency ratio (coefficient of performance) of the refrigerating unit can reach more than 5, achieving high-efficiency refrigeration, reducing refrigeration cost and improving economic benefits. During the heating link, the PV thermoelectric module generates low-temperature heat energy while generating electricity, and makes full use of this part of heat that shall be discarded as one of the energy sources for heating or domestic hot water. Meanwhile, the air source heat pump is used as the standby heating source to ensure the heating stability under adverse weather conditions. This technology can effectively improve the primary energy utilization rate of the whole functional system, realize the cascade utilization of energy, and effectively improve the economy and comprehensive benefits of the energy supply system.

Table 3-3   New equipment electricity consumption statistics for long-term scenarios

Energy using division	Equipment	Average power	Quantity	Average daily electricity consumption	Daily operation mode
Technology Demonstration Center	Rice processing	8.56 kW	8	684.8 kWh	Operate for 10h in the daytime
	Chili powder production	24 kW	1	240 kWh	Operate for 10h in the daytime
	Bread production	6.52 kW	8	521.6 kWh	Operate for 10h in the daytime
	Training classroom Electric equipment	180 kW	1	1440 kWh	Operate for 8h in the daytime
	Street lamp	40 W	59	18.88 kWh	Operate for 8h at night
Total		—	—	2905.28 kWh	—

Design scheme

◆ Power generation combination design of near–term scheme

Off-grid PV power generation system mainly includes PV power generation unit, energy storage unit and distribution unit. Based on the power load operation characteristics of agricultural park in the current phase, the design ratio of PV and energy storage is carried out according to the local solar energy resource endowment and the principle of power balance and electricity balance. The configuration scheme is shown in Table 3-4 below.

Table 3-4   Configuration of PV power generation system

S/N	Item	Technical specifications	Quantity	Unit	Total price/ RMB10,000
1	PV power generation unit	250 kW	1	Set	100
2	Energy storage unit	140 kW/280 kWh	1	Set	46.2
3	Distribution unit	1 low voltage distribution cabinet	1	Set	5
4	International freight	—	1	Item	10
5	Construction cost	—	1	Item	10
6	Project management fee	—	1	Item	10
Total		—	—	—	181.2

field measurement of agricultural park staff. Among them, the daytime electricity consumption is about 684.4kWh, while the night electricity consumption is about 226.8kWh. Table 3-2 lists the electricity consumption statistics of newly added devices in the near-term scenario.

Table 3-2   Equipment electricity consumption statistics for near-term scenarios

Energy using division	Equipment	Average power	Quantity	Average daily electricity consumption	Daily operation mode
Living and office area	Water heater	1.5 kW	31	46.5 kWh	Operate for 1h at night
	Street lamp	100 W	42	42 kWh	Operate for 10h at night
	Air conditioner	1.5 kW	20	360 kWh	10 sets operate full-time on average
	Freezer/refrigerator	200 W	13	62.4 kWh	Full-time operation
	Illuminating lamp	20 W	183	18.3 kWh	Operate for 5h at night
	Water dispenser	420 W	8	6 kWh	Random operate < 2h
	Disinfecting cabinet	800 W	4	16 kWh	Random operate for 5h
Oil extraction plant	Oil extraction equipment	45 kW	1	360 kWh	Operate for 8h in the daytime
Total		—	—	911.2 kWh	—

As for the long-term scheme, the energy demand of agricultural park shall meet the energy demand after the technology demonstration center is put into operation on the basis of the near-term scheme. The center contains 8 set of rice processing and production equipment, 1 set of chili powder production equipment, 8 set of bread production equipment, 1 set of electrical equipment for training classrooms and 59 street lamps. The total power of the above equipment is about 327kW, and the total power of the long-term scheme equipment is about 463.72kW. After calculation, the average daily electricity consumption of the long-term scheme is about 3816.48kWh. Among them, the daytime electricity consumption is about 3570.8kWh(911.2kWh in near-term+new increase of 2905.28kWh in long-term), while the night electricity consumption is 245.68kWh(226.8kWh in near-term+new increase of 18.88kWh in long-term). Table 3-3 lists the electricity consumption statistics of newly added devices in the long-term scenario.

current production and living status quo of the agricultural park and the long-term development needs. The near-term upgrading mainly meets the current production and living energy demand of agricultural park, including living and office areas and oil extraction plants, etc. By introducing PV power generation, an off-grid power supply system is built to solve the problem of current power shortage and economy in the agricultural park. The long-term upgrading mainly focuses on the production and living energy demand after the technology demonstration center is put into operation, including: living and office areas, oil extraction plants and technology demonstration centers, etc. By building PV off-grid power generation systems and introducing "new energy + agriculture" off-grid equipment under multiple scenarios, we can meet the electricity demand of agricultural park and create a green, economical, low-carbon, environmentally friendly and advanced model of modern agricultural park.Table 3-1 shows the near-term and long-term scenarios for off-grid power supply transformation.

Table 3-1   Near-term and long-term scenarios of off-grid power supply transformation

Design scenario	Design orientation	Design purpose
Near-term scheme	Meet the current energy demand for production and living, including living and office areas and oil extraction plants	Solve the problem of current power shortage and economy in the agricultural park
Long-term scheme	Meet the energy demand after the technology demonstration center is put into operation on the basis of meeting the current energy demand for production and living; introduce "new energy + agriculture" off-grid equipment under multiple scenarios at the same time	Create a model of green, economical, low-carbon, environmentally friendly and advanced modern agricultural park

Electricity consumption analysis

As for the near-term scheme, the energy-using of the agricultural park mainly includes the living and office areas and the oil extraction plants, in which the living and office areas contain 31 water heaters, 42 street lamps, 20 air conditioners, 13 freezers/refrigerators, 183 illuminating lamps, 8 water dispensers and 4 disinfection cabinets; the oil extraction plants contains 1 set of oil extraction equipment. The total power of the above-mentioned equipment is about 136.72kW, and the average daily electricity consumption is about 911.2kWh according to the

3.The sustainable, stable and environment-friendly energy supply can promote the green, environment-friendly and low carbon emission construction of the park.

4.The PV new energy with low power generation cost replaces the power generation with high input, which can improve economic benefits of the industrial park.

5.A typical way of "new energy + agriculture" application is adopted to radiate Africa and open up the development of African agricultural innovative economic model.

Design idea

According to the investigation of renewable energy resource endowment in the project site, the local solar energy resources are abundant, and the annual total radiation amount is $1889.8kWh/m^2$; the wind resource is poor, the wind speed at 50m height is about 3.8m/s, and the wind power density is about $60W/m^2$. According to the calculation, the PV power supply project has the best economy. Therefore, the off-grid upgrading of agricultural park takes PV as power supply and combines electrochemical energy storage to build a new energy off-grid power generation system (Fig.3-7).

Fig. 3-7 PV energy storage off-grid power generation system

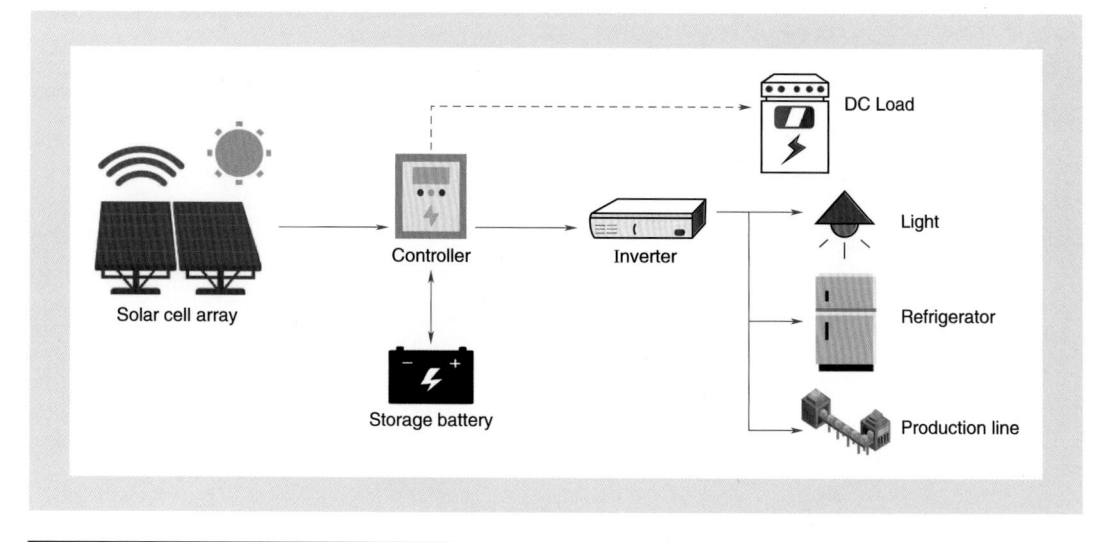

According to the idea of "overall planning, far and near integration and benefit priority", the off-grid power supply upgrading and transformation is conceptually designed according to two scenarios of "near-term plan" and "long-term plan" on the basis of

3.2 CONCEPTUAL DESIGN OF OFF-GRID TRANSFORMATION OF THE AGRICULTURAL PARK

Design purpose

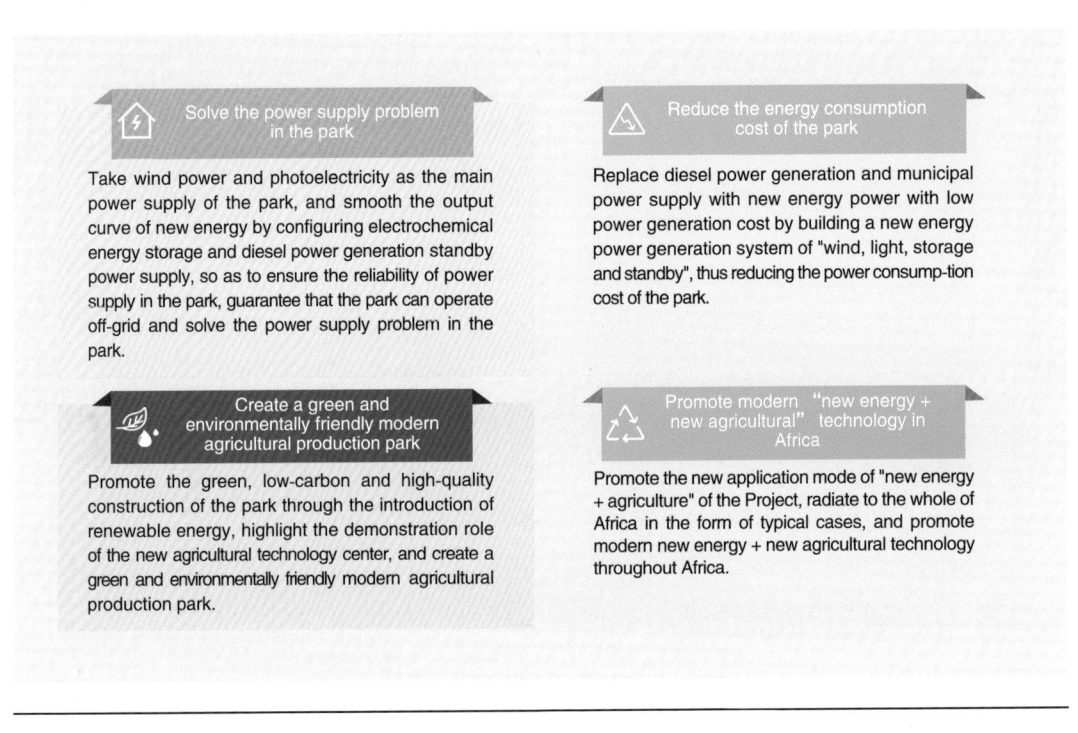

Solve the power supply problem in the park

Take wind power and photoelectricity as the main power supply of the park, and smooth the output curve of new energy by configuring electrochemical energy storage and diesel power generation standby power supply, so as to ensure the reliability of power supply in the park, guarantee that the park can operate off-grid and solve the power supply problem in the park.

Reduce the energy consumption cost of the park

Replace diesel power generation and municipal power supply with new energy power with low power generation cost by building a new energy power generation system of "wind, light, storage and standby", thus reducing the power consump-tion cost of the park.

Create a green and environmentally friendly modern agricultural production park

Promote the green, low-carbon and high-quality construction of the park through the introduction of renewable energy, highlight the demonstration role of the new agricultural technology center, and create a green and environmentally friendly modern agricultural production park.

Promote modern "new energy + new agricultural" technology in Africa

Promote the new application mode of "new energy + agriculture" of the Project, radiate to the whole of Africa in the form of typical cases, and promote modern new energy + new agricultural technology throughout Africa.

Design principle

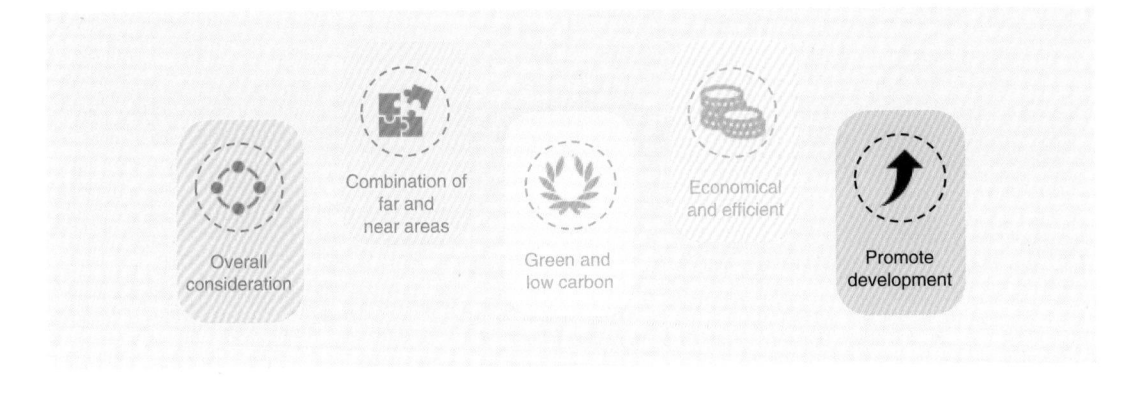

Overall consideration

Combination of far and near areas

Green and low carbon

Economical and efficient

Promote development

1.It not only meets the power supply needs of the park and reduces the cost of electricity, but also reflects the exemplary role to build a green agricultural park.

2.The near-term plan is adopted to solve current energy demand, while the long-term plan is used to build a green "new energy + agricultural" application model.

Now it has become a well-known multi-functional PV agricultural industrial base in the local area and a demonstration park where PV industry drives the comprehensive development of aquaculture and adjacent industries.

PV + grain processing

The most common application mode of "PV + grain processing" is to lay PV panels

on the roof of granaries with a large area of land. PV panels on the roof of granaries have the effect of heat preservation and insulation, which improves the grain storage environment inside granaries. Energy is generated by PV power generation on the roof of granary, which provides power for operating devices such as temperature control device and humidity control device in grain storage system, realizes sustainable development of low-temperature grain storage in granary, saves operating cost and reduces resource waste, and has good economic benefits.

Case: PV power station on the roof of granary

The rooftop PV power station of Caoban National Granary in Zhangzhou City is located in Jingcheng Town, Nanjing County, northwest of Zhangzhou City. In the Project, the standardized warehouse roof with a total area of $20000m^2$ in Caoban Granary is used to build PV power generation modules with a total installed capacity of 2MW. Part of the electric energy generated by PV power generation is used to keep the operation of temperature control system and humidity control system in the granary, and the rest of the electricity can be connected to the local power grid. In the project, it makes full use of the superior resources of the whole city's grain depot roof to realize the effective integration of economic benefits and scientific and technological grain storage.

3.1 PROJECT OVERVIEW — AGRICULTURAL PARK IN ABUJA

drought-resistant system in China. The PV panel covers an area of 77m², with an output power of 7680W and a water pumping lift of over 100m. In the dry period, the electric energy generated by solar energy is used in the project to pump water and fight drought and irrigate farmland, and the surplus electricity is sent into the local power grid in the non-dry period, so as to obtain the extra income of the project. The system solves the problem of farmland irrigation in the dry period and increases the local income.

PV + aquaculture

"PV + aquaculture" is used to build a PV power generation system above aquaculture waters, carry out PV power generation on water and fishery culture underwater. This mode is called fishing-light complementation PV power station. The mode of "fishing-light complementation" can not only make full use of space and save land resources, but also use PV power stations to adjust the breeding environment, reduce water evaporation and inhibit blue-green algae reproduction, and improve the economic value and water environment per unit area of water, with good economic benefits and environmental effectiveness.

Case: fishing–light complementation PV power station

The installed capacity of "fishing-light complementation" PV power generation project in Xinghua City, Jiangsu Province is 178MW, located in Shagou Town, northwest of Xinghua City. PV modules are erected above fish ponds and wetlands to generate electricity to form the mode of "generating electricity at the upper part and raising fish at the lower part", which not only makes full use of space, saves land resources, but also adjusts the breeding environment by using PV power stations.

Case: Cowshed Distributed PV Power Station

Cowshed Distributed PV Power Station in Cixi City, Zhejiang Province, China is located in Cixi Modern Agricultura Development Zone, Zhejiang Province, covering an area of 0.8km^2 with an installed capacity of 15.6MW. In the Project, PV panels are laid on the roof of the newly-built cowshed to generate electricity while shielding sunlight in summer, creating a comfortable internal environment for livestock and poultry in the shed. The electric energy generated by PV power generation can be used as the power source for environmental control in the shed. All the power generation by the project is merged into the local power grid, which brings good economic benefits and comprehensive benefits of land use to users.

PV + irrigation by pumping

"PV + irrigation by pumping" is mainly composed of solar cell array, inverting control system and water pumping and conveyance system. Under the condition of

sufficient sunshine in dry season, the unfavorable conditions in agricultural production are turned into advantages. PV power generation is used to 3 drive water pumps, pumping units and other water conservancy machinery to pump and convey water to realize drought-resistant irrigation. This PY operation mode has been implemented in many areas with water and no electricity, which has solved the farmland irrigation problem in these areas and guaranteed the vital interests of local people.

Case: Solar water pumping and drought–resistant wheat watering system

The installed capacity of 7kW solar water pumping and drought-resistant wheat watering system in Henan Province, China is the first solar water pumping and

"PV + field planting" is a land use method to address the contradiction between agricultural land and production land, which achieves multi-purpose use of land and improves the economic benefits and utilization efficiency of the land.

Case: PV power generation leading base

The national PV power generation leading base in Tongchuan City, Shanxi Province, China has an installed capacity of 25MW and covers an area of about 6.33km^2. The design orientation is "PV power generation + agricultural planting + sightseeing tour + poverty alleviation". In addition to its advantage of meeting the electricity demand of surrounding areas and the energy supply of greenhouses, medicinal materials can be planted under PV panels. The power station not only provides green power for local production and life, but also promotes a new mode of agricultural planting. On the basis of creating a large number of employment opportunities, it promotes local people to get rid of poverty and increase income, and plays an active role in ecological restoration of the Loess Plateau at the same time.

PV + livestock and poultry breeding

"PV + livestock and poultry breeding" includes indoor breeding in livestock houses and PV open-air breeding Among them, the electric energy generated by PV power generation is used in indoor breeding in livestock houses to regulate the internal environment of livestock houses, such as temperature, humidity and exhaust devices; the combination of PV system and open-air breeding of livestock and poultry means planting pasture under PV panels which provides sufficient nutrients for livestock and poultry and "shade" for livestock and poultry  at the same time so as to avoid excessive exposure and burn of livestock and poultry. This combination of PV power generation and livestock and poultry breeding has considerable energy efficiency, and it is also efficient utilization of agricultural land resources.

Chinese "carbon peaking and carbon neutrality" put forward, the green and low-carbon transition of economy and society, which is mainly characterized by clean energy system, green industrial structure, electrified consumption mode and low-carbon lifestyle, will be accelerated in an all-round way, and the "new energy + agriculture" model will also usher in the fast development, promoting the realization of high-quality development of green, low-carbon and circular agriculture.

Fig. 3-6   Common application scenarios of "new energy + agriculture" in China

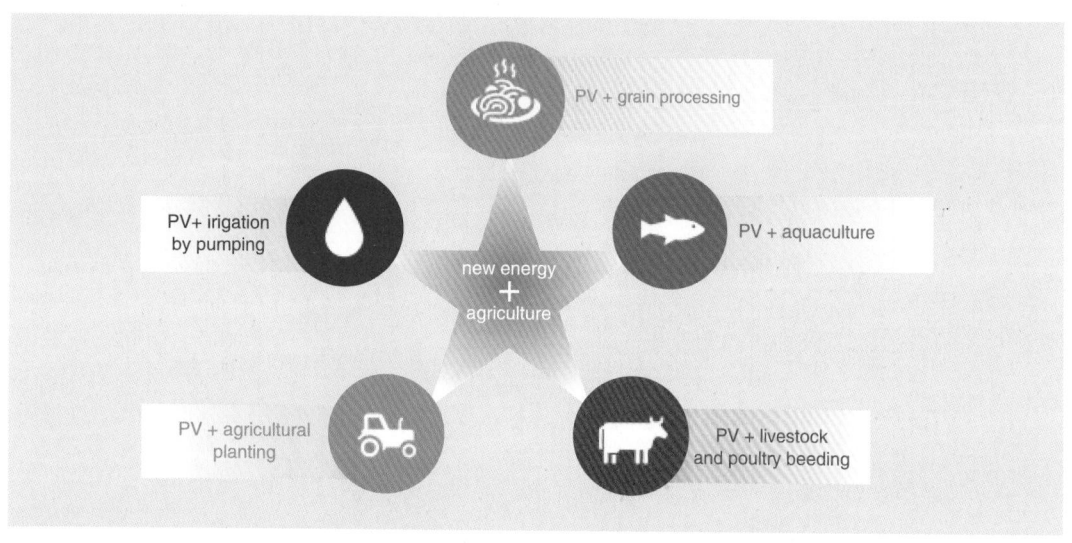

PV + agricultural planting

"PV + agricultural planting" is the most widely used, mainly including two modes of "PV + facility gardening" and "PV + field planting". Among them, "PV + facility

gardening" combines PV generation with greenhouse cultivation. Under the conditions of meeting the light requirements for crop growth, solar panels are installed on the top of the greenhouse to use PV generation to meet the electricity needs inside the greenhouse. By fully utilizing the advantages of light control, temperature control, and pest prevention in the greenhouse, high-quality products are produced.

3.1   PROJECT OVERVIEW — AGRICULTURAL PARK IN ABUJA

the agricultural park is planning to introduce new energy equipment and establish a self-sufficient small off-grid power supply system to provide clean power for the agricultural park, improve the reliability and economy of power supply so as to build the park into a modern "new energy + agriculture" demonstration project.

COLUMN: APPLICATION OF "NEW ENERGY + AGRICULTURE"

"New energy + agriculture" is one of the important directions of modern agricultural development. New energy is clean, low-carbon, environmentally friendly, highly economical, renewable, locally available and flexibly arranged especially suitable for the strong dispersion of agricultural population in Africa. Adopting the modern technology of "new energy + agriculture" cannot only realize stable energy supply under various agricultural production scenarios in remote areas, but also carry out the flexible arrangement of new energy equipment according to local conditions, realizing the mode innovation of agricultural production, promoting the effective complementarity between new energy technology and agricultural production, effectively improving the agricultural production efficiency in remote areas without access to electricity, and driving the local economic and social development.

Fig. 3-5   Four advantages of application of "new energy + agriculture" in Africa

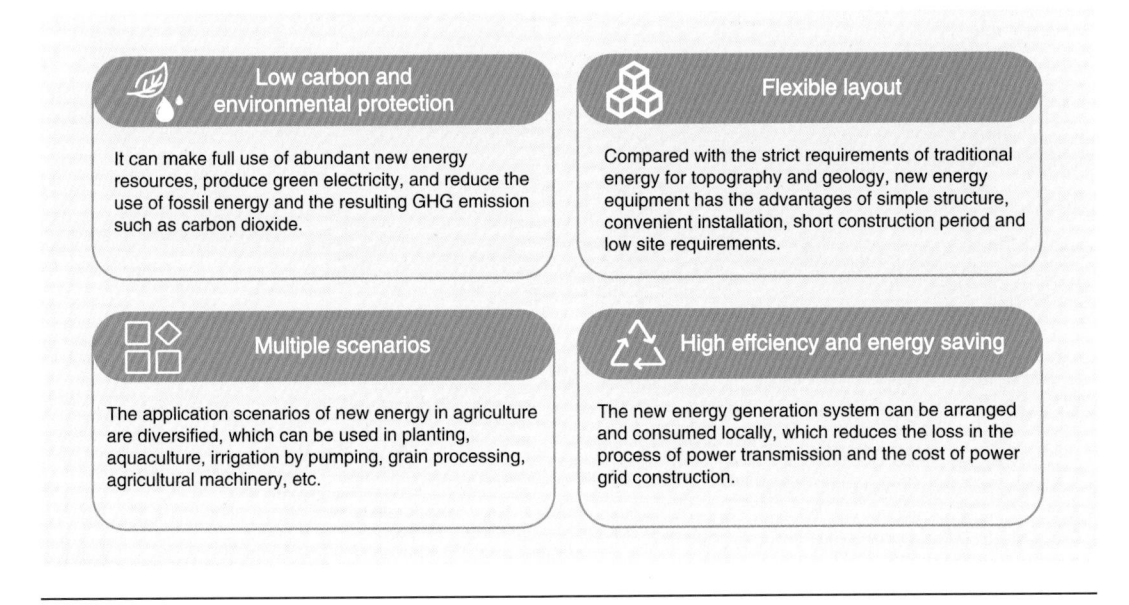

The increase of agricultural power demand and the decrease of new energy power generation cost promote the large-scale application of new energy in agriculture. China is a great agricultural country and a great energy country in the world. With the goal of

Status quo of electricity consumption in agricultural park

For the agricultural park, the annual electricity consumption is about 323 MWh, the total power of various consumers is 879.5kW, and the peak value of daily power load usually occurs in the daytime and is about 80~90kW. The energy consumption diagram of the agricultural park is shown in Fig. 3-4, the agricultural park mainly includes two main electricity consumption areas: living and office areas and oil extraction plants."The living and office area includes the dormitory area and the public service area and the peak power load in the dormitory area is about 31kW. The public service area is composed of kitchen, dining room and laundry room and has the peak power load of about 10kW; the peak power load of the oil mill is about 45kW.

Fig. 3-4   Schematic diagram of energy consumption of the agricultural park

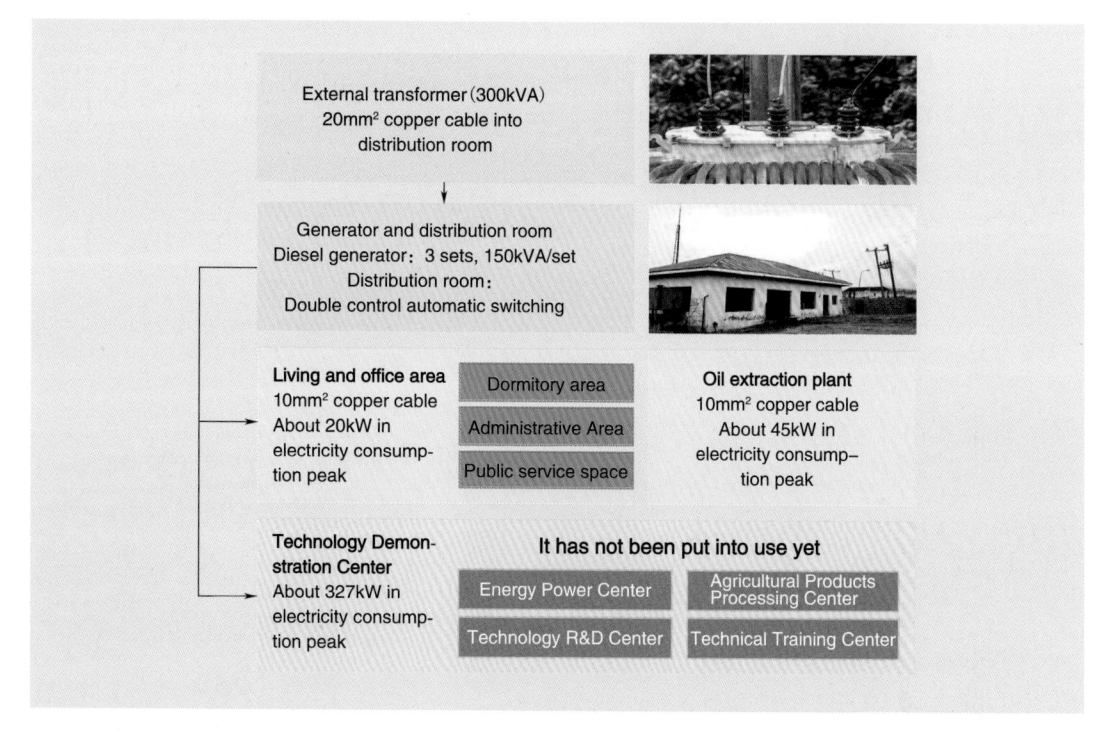

At present, the agricultural park is mainly powered by municipal power supply introduced by a 300kVA external power transformer. Due to frequent power failure in the agricultural park, the quality of municipal power supply cannot be guaranteed. Therefore, three 150kVA diesel generators have been procured for the agricultural park and can be used as standby power supplies alternately. Their monthly diesel fuel consumption is about 4000L. Along with the rising price of diesel fuel, the cost of electricity has risen significantly in the park. Therefore,

Functional zoning of agricultural park

As shown in Fig. 3-3, the agricultural park currently includes living and office area, oil extraction plant, etc. The living and office area is located in the northeast of the agricultural park and has main functions as daily office work, living services, etc.; The main function of the oil extraction plant is to refine edible oil. The agricultural park will also bring into play a technology demonstration center in the future, which will be composed of energy power center, agricultural product processing center, technology research and development center, technical training center, etc. and have main functions as power generation and energy supply, agricultural product processing, seed and seedling research and development, agricultural machine training, etc.

Fig. 3-3 Schematic diagram of functional partition of the agricultural park

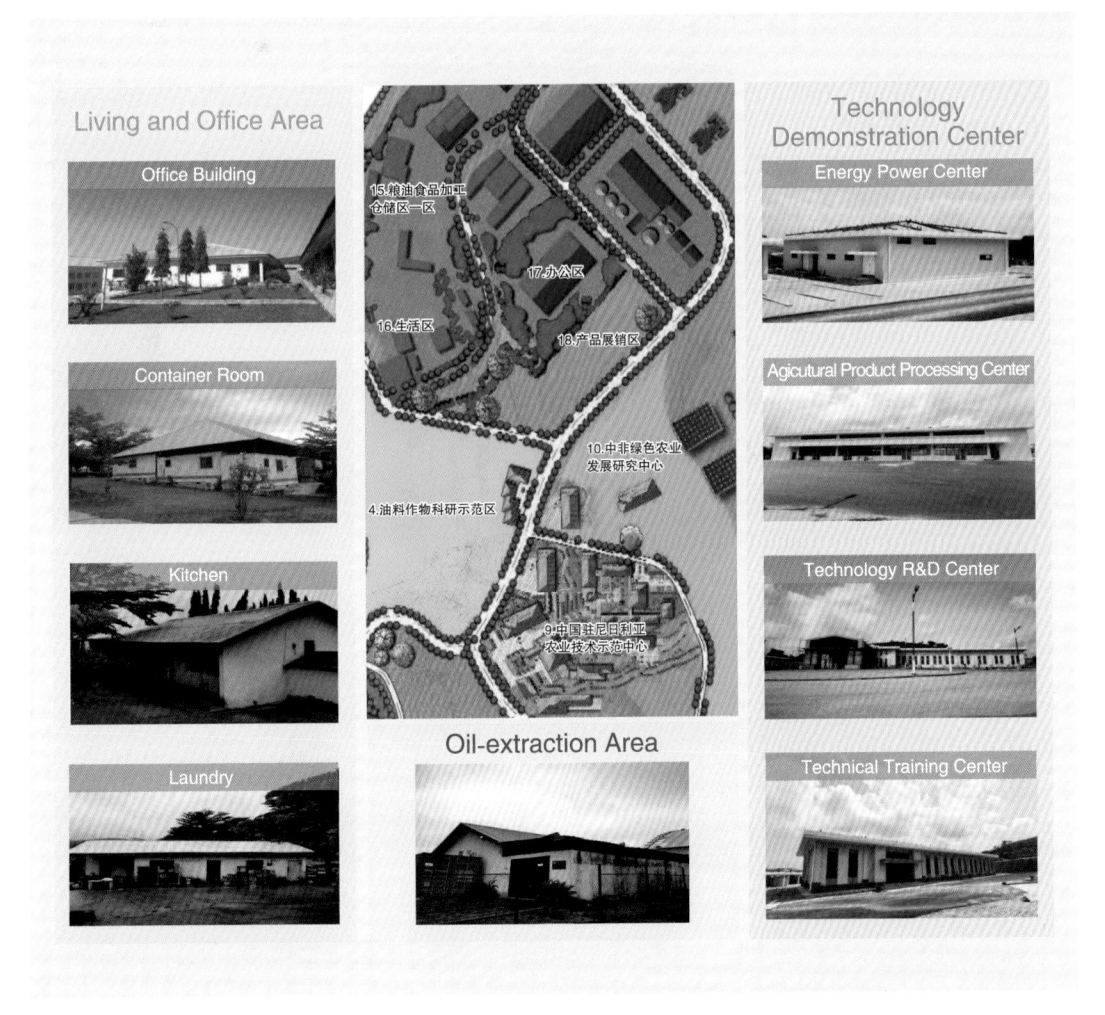

Chinese agricultural culture and civilization and Chinese agricultural technology transfer center in Africa; it is a cooperative platform for the academic exchange of knowledge and information and the exhibition and trading of elements, products and resources to promote agricultural development in Africa.

Fig. 3-1   Schematic diagram of geographical location of the agricultural park

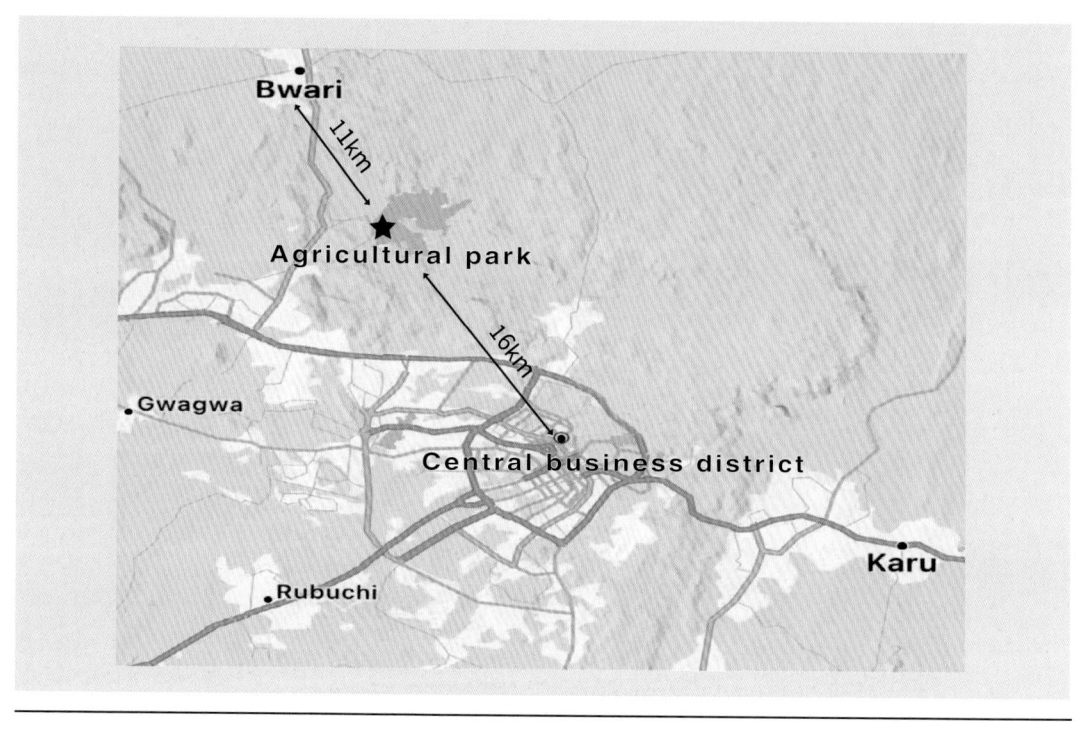

Development orientation of industrial park

As shown in Fig. 3-2, the development positioning of the agricultural park includes the Agricultural Exhibition and Trading Cooperation Platform, the Africa-China Agricultural Technology Transfer Center, and the Modern Agricultural and Pastoral Complex.

Fig. 3-2   Development positioning of the agricultural park

Agriculture is an important pillar industry for African economic and social development and is the main source of income for rural population in most places of Africa. According to the data of the Food and Agriculture Organization (FAO) of the United Nations, the agricultural employment population in Sub-Saharan Africa has accounted for more than 50% of total labor force in Africa. The average contribution rate of agriculture to GDP of African countries is 15%, while the global average value is only 4.1%. The African continent has about 930 million hm^2 of land suitable for agricultural production, which is equivalent to the land area of the United States.

However, in many regions in Africa, traditional farming equipment are still used due to economic conditions and lack of reliable power access. So, the huge potential of African agricultural development is far from being released. Providing production energy for remote rural areas in Africa through off-grid can effectively improve agricultural production time and efficiency, which is of great practical significance for promoting economic and social development in Africa. In order to make African off-grid projects more representative, this study selects and researches an off-grid agricultural project, which is valuable for promotion - CGCOC Agriculture Abuja High-Tech Industrial Park (hereinafter referred to as "Agricultural Park").

3.1 PROJECT OVERVIEW — AGRICULTURAL PARK IN ABUJA

Basic overview of agricultural park

As shown in Fig. 3-1, the agricultural park is located in Ushafa Village, Buwari District, northwest of the central business district of Abuja, the capital of Nigeria, is about 11km away from the north outer ring highway of the capital and covers a total area of 89 hm^2. The agricultural park is adjacent to hills on one side and highways on the other three sides and is next to the reservoir at the lower reach of the river. So, it has good traffic, water, labor and social security conditions. The agricultural park is the first modern agricultural demonstration comprehensive park in West Africa and its main work includes seed and seedling research and development, agricultural product processing, horticultural cultivation, agricultural material sales, agricultural machine training and scientific research office. The park is oriented as an industrial incubation park radiating Africa with Nigeria as the center; it embodies the agricultural pastoral complex of

3

CASE STUDY OF OFF-GRID PROJECTS IN AFRICA

Taking the Agricultural High-tech Industrial Park in Abuja as the object, by studying the current electricity consumption characteristics in combination with the new modern agricultural concept of "New Energy + Agriculture", we aim to put forward conceptual design for the off-grid upgrading and transformation plan for the park's stable power supply and energy consumption cost control. In addition the plan's economic and social benefits will be studied to provide reference programs and technical ideas for promoting the "New Energy + Agriculture" concept in Africa.

Selectively implement the outstanding full–subsidy projects

The government should provide certain financial subsidies for early construction and later operation of off-grid projects in order to protect basic electricity rights and interests enjoyed by low-income groups, solve the electricity access of rural people, who are living in remote areas in Africa, have urgent need of electricity demand and have low electricity affordability, and promote the access to electricity in Africa. Therefore, it is recommended to selectively implement the best full-subsidy projects. It is required to focus on the projects, which can benefit residents in remote villages and towns, is of great significance in universality and demonstration and can meet the most basic electricity demand of low-income people without access to electricity during project planning.

Strategies for off-grid project development

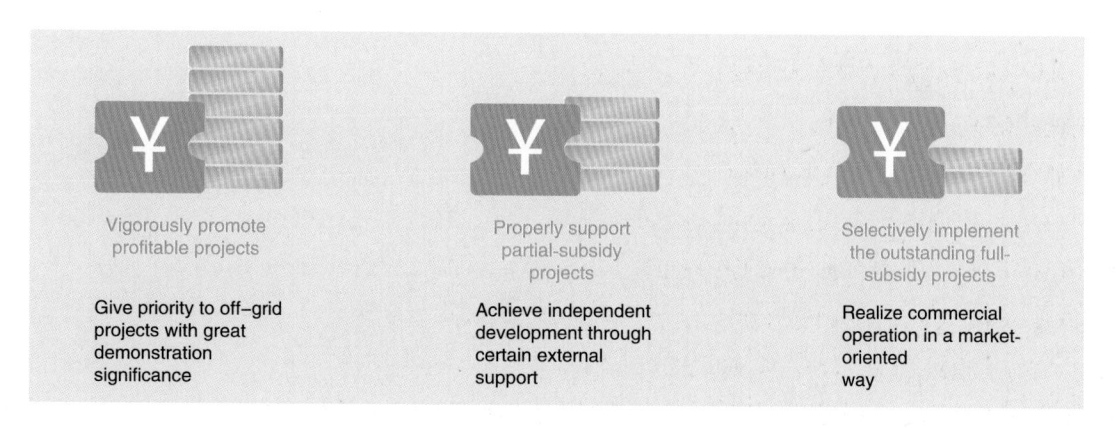

Vigorously promote profitable projects

Give priority to off-grid projects with great demonstration significance

Properly support partial-subsidy projects

Achieve independent development through certain external support

Selectively implement the outstanding full-subsidy projects

Realize commercial operation in a market-oriented way

Vigorously promote profitable projects

In order to guarantee sound development of the off-grid market, government intervention and financial expenditure should be reduced as much as possible and the sustainable operation of the project should be preferentially achieved by the market-oriented approach. Therefore, it is recommended to vigorously promote profitable projects in terms of development strategies. During project planning, priority should be given to whether there are Type I users with high electricity demand and stable affordability in the region. The project should be preferentially distributed near industrial and mining enterprises and other high electricity consumption areas, concentrated high-quality industrial areas as well as areas with favorable policies to fundamentally ensure the economy of off-grid projects.

Properly support partial-subsidy projects

In order to benefit more people's livelihood and meet the electricity demand of users with low affordability, the government should provide certain financial subsidies for early construction of the off-grid projects, so that the projects can achieve independent development in the later stage. Therefore, it is recommended to properly support partial-subsidy projects. During project planning, it is required to note whether there are Type II users of a certain scale in the region and whether the electricity charge can cover the later operation costs. The project should be preferentially distributed near businesses engaged in production and business activities and with concentrated electricity consumption, so as to promote the balanced development of the regional economy.

O&M is operation and maintenance expense; *ret* is the cost of selling electricity;

F_t is the fuel cost in Year t;

E_t is the power generation in Year t;

r is the discount rate;

n is the financial calculation period.

COLUMN 2: COMPREHENSIVE AFFORDABLE ELECTRICITY PRICE FOR USERS

The comprehensive affordable price for users determines the upper limit of the average affordable price on the user side and is the basis for rational formulation of electricity policy. Generally, the comprehensive affordable electricity price for users is determined through on-site investigation and through analysis and calculation of the per capita disposable income, the proportion of electricity charges in disposable income, electricity consumption and other data of various typical users on the user side.

2.4 STRATEGIES FOR OFF-GRID PROJECT DEVELOPMENT

The off-grid project should be developed from the actual local energy demand and the project development mode should be scientifically determined according to the electricity affordability of user groups and the actual situation of the project. In order to promote the development of off-grid projects while government subsidies is reduced as much as possible, so as to ensure the steady and sustainable development of the market and continuously benefit the population without access to electricity in remote areas, the competent government departments should formulate different project development strategies for three types of projects and develop corresponding incentive measures on the basis of the principles of "scientific planning, adaptation to local condition and gradual progress". It is recommended to vigorously promote profitable projects and give priority to the principle of market commercialization to solve the problem of electricity access for people without access to electricity; properly support partial-subsidy projects, and achieve independent development and sustainable operation of the project through a certain degree of political support; preferentially carry out the fully-subsidy projects. For livelihood projects with urgent needs, outstanding demonstration significance and obvious benefits, but without commercial operation conditions, the non-market-oriented subsidy methods should be considered for support.

When $\overline{C} \geqslant LCOE$, There is no external subsidy and the profit model is adopted in the project.

When $\overline{C} < LCOE$, if $P_{O\&M} \leqslant \overline{C} < P_l$, The electricity revenue can cover the operation and maintenance costs in the later stage of the project, but cannot cover the capital costs (including equity costs) in the earlier stage of the project. So, the partial subsidy model is adopted for the project.

When $\overline{C} < LCOE$, if $\overline{C} < P_{O\&M}$, The electricity revenue can cover neither the capital cost (including equity cost) in the early stage of the project, but also the operation and maintenance cost in the later stage of the project. So, the full subsidy mode is adopted for the project.

COLUMN 1: LEVELIZED COST OF ENERGY (LCOE)

LCOE is a short form of "Levelized Cost of Energy". As a quantitative economic indicator, LCOE is often used to compare and evaluate the comprehensive economic benefits of energy and power generation projects. It is the power generation cost calculated after the cost and power generation are averaged within the life cycle of the project, namely the ratio of the present value of the cost to the present value of the power generation within the whole life cycle of the project. LCOE is widely used in the world to measure the relationship between benefits and costs of different energy and power generation projects and to provide references for formulating electricity selling price and scheme.

The calculation formula of levelized cost of energy (IRENA, 2018) is as follows

$$LCOE = \frac{\sum_{t=1}^{n} \dfrac{I_t + M_t + F_t}{(1+r)^t}}{\sum_{t=1}^{n} \dfrac{E_t}{(1+r)^t}} \tag{2-1}$$

$$M_t = f(dep, fin, O\&M, ret)$$

where,

$LCOE$ is the levelized cost of energy;

I_t is the investment in Year t (including investment return);

M_t is the operating cost in Year t, where *dep* is depreciation expense; *fin* is financial expense;

2.3   JUDGMENT OF DEVELOPMENT MODELS OF OFF-GRID PROJECTS

Table 2-2 Classification of off-grid development models

Development model	Purpose	Subsidy method	Characteristics of model
Profit model	Profit making	No subsidy	The electricity revenue can fully cover the capital cost (including equity cost) in the early stage and the operation and maintenance cost in the later stage of the project, which can be put into commercial operation without external subsidies
Partial subsidy model	Profit+poverty alleviation	Subsidy forearly stage Capital cost	The capital cost in the early stage of the project will be externallysubsidized once, and the operation and maintenance cost will be covered by the electricity revenue in the later stage, so as to realize the sustainable operation of the project
Full subsidy model	Poverty alleviation	Whole process subsidy	The electricity revenue cannot cover the operation and maintenance cost in the later period, so that the external subsidy must be provided throughout the process to realize the sustainable operation of the project

2.3 JUDGMENT OF DEVELOPMENT MODELS OF OFF-GRID PROJECTS

According to Table 2-2, the judgment of Type Ⅲ development models of off-grid projects is closely related to the extent to which the electricity revenue covers the full life cycle costs (including equity costs) of the project. The electricity revenue depends on the comprehensive affordable electricity price $(\bar{C})$ on the user side, and the full life cycle cost (including equity cost) of the project can be directly reflected by the levelized cost of energy (LCOE). The judgment methods for development models of off-grid projects are introduced as follows:

Fig. 2-1 Flowchart for determination of development models

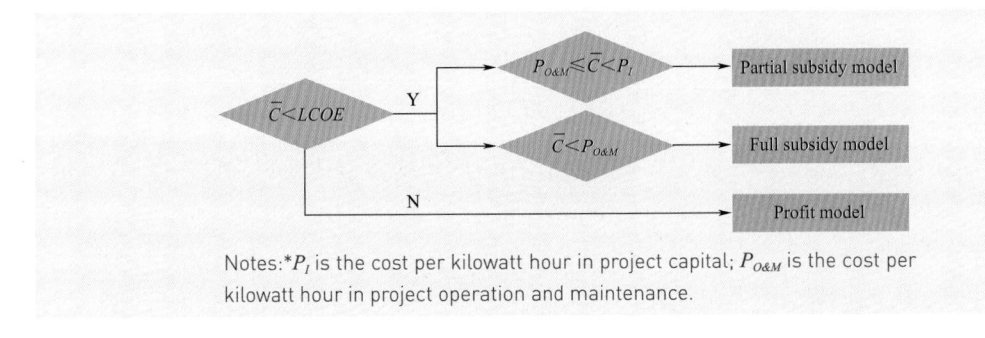

Notes:*P_I is the cost per kilowatt hour in project capital; $P_{O\&M}$ is the cost per kilowatt hour in project operation and maintenance.

Table 2-1 Division of off-grid user groups

User type	Electricity demand	Affordability	Main function	User composition
Type I	Fairly high	Stable	Guarantee project economy to a large extent	Mine, metallurgy, production workshop, etc.
Type II	Moderate	Relatively stable	Guarantee project economy to a certain extent	Bazaar, vegetable market, hospital, school, commercial building, etc.
Type III	Smaller	Unstable	Need electricity charge supplement, unable to guarantee the economy of the project	Common families, self-employed households, etc.

2.2 CLASSIFICATION OF DEVELOPMENT MODELS OF OFF-GRID PROJECTS

The classification of off-grid project development models should be scientifically judged according to local actual situation. Due to different distribution of various user groups in different regions, the electricity revenue may not fully meet the requirements of pure commercial operation of the project and the sustainable operation of some projects have to rely on government or other external subsidies. According to external subsidies needed or not and the diferent ways of subsidies, the off-arid project development modes can be divided into profit model, partial subsidy model and full subsidy model.

As shown in Table 2-2, the profit model follows the market business principle, the electricity revenue can fully cover the capital cost (including equity cost) in the early stage and the operation and maintenance cost in the later stage ofthe project; the partial subsidy model does not take profit as the primary purpose, but has the dual attributes of profit and poverty alleviation, the electricity revenue can cover the operation and maintenance costs of the project in the later stage, but the capital costs (including equity costs) in the early stage of the project have to rely on government or other external subsidies in whole or in large part the full subsidy model is not for profit and has the nature of poverty alleviation. The electricity revenue can neither cover the capital cost (including equity cost) in the early stage of the project nor the operation and maintenance cost in the later stage of the project. The project construction depends entirely on grants, subsidies or preferential loans provided by the government or outside.

The users of off-grid projects are mostly distributed in remote areas. The economic foundation is weak in remote areas in Africa, where the affordability of electricity prices of local residents is weak. It is difficult to realize the economic sustainable operation of off-grid projects only by electricity revenue of local residents. Therefore, it is necessary to analyze the development model of off-grid projects.

2.1 DIVISION OF USER GROUPS FOR OFF-GRID PROJECTS

The user groups of projects should be classified in order to guarantee the economic and sustainable operation of off-grid projects. Through reasonable allocation of different user groups, the economy of the project is guaranteed to a certain extent on the one hand, and the electricity price is shared under the principle of fairness and reasonableness on the other hand. Through cross subsidies of high-income groups to low-income groups, it is guaranteed that low-income groups can enjoy the inclusive basic power rights and interests.

According to the power demand and payment capacity of off-grid projects, the user groups are divided into three categories, namely, Type I, Type II and Type III, as shown in Table 2-1. Type I of users refers to those with high electricity demand and stable affordability. They can ensure sustainable economic operation of the project to a large extent and are generally large- and medium-sized industrial users, such as factories, mines and production workshops of a certain scale. Type II of users refers to those with moderate electricity demand and stable affordability. They can ensure the economic sustainable operation of the project to a certain extent and are generally commercial users, such as bazaars, vegetable farms, hospitals, schools, commercial buildings, etc. Type III of users refers to those with small electricity demand and low affordability. They can only be used as a supplement to the electricity charges of off-grid projects and cannot guarantee the economic sustainable operation of the project, such as ordinary families, self-employed households, etc.

2

ANALYSIS ON BUSINESS MODEL OF OFF-GRID PROJECTS IN AFRICA

Guided by the sustainable operation of off-grid projects, this chapter classifies the development patterns in view that whether external financial subsidies are needed and how the subsidies are used. By comparing the users' electricity price tolerance with the levelized cost of energy (LCOE) of the project, this chapter proposes methods to choose off-grid development patterns, as well as subsidize, so as to realize sustainable operation of off-grid projects when groups with insufficient access to electricity enjoy inclusive electricity consumption rights and interests.

solar home system and is one of top three countries using this technology in Africa.

6. Mozambique

The Mozambican Government approved the Off-grid Energy Access Regulation (ROGEA) in December 2021 to provide legal framework support for the off-grid field, increase the transparency of relevant participants, provide necessary conditions for the private sector, and protect investment in various off-grid technologies, e.g. solar home systems and microgrids. Promoting off-grid electrification is an important part of Mozambique's Energy for All (MEFA) Project in 2021. It is planned to achieve 100% access to electricity nationwide by 2030, of which 30% will be provided by off-grid power generation. In June 2022, Mozambique's National Energy Fund (FUNAE) received USD 26 million from the World Bank to promote off-grid development and solar home system (SHS) projects, which are expected to benefit about 300,000 people in the next 4 years.

7. Niger

The Niger Government reached an agreement with the International Development Association (IDA) of the World Bank in 2017, which plans to provide USD 50.3 million to support Niger Solar Electricity Access Project (NESPA), aiming to realize the electrification of rural areas in Niger through solar energy. Main contents of NESPA include market development of independent PV equipment, rural electrification of solar microgrid, isolated solar hybrid microgrid system, implementation support and technical assistance. The project will benefit 352 villages and promote the electrification of rural areas in Niger.

8. Rwanda

According to the 7-year government plan: National Transformation Strategy (NST1, 2017–2024) formulated by the Rwandan Government, the off-grid power supply is regarded as an important part of national electrification and the development goals are proposed in terms of power generation, power supply quality and reliability, power supply capacity, etc. It is planned to achieve 100% access to electricity in the country in the fastest and most economical way by 2024, of which the proportion of main grid power supply is 52% and the proportion of off-grid power supply is 48%. The off-grid technology has gradually matured in recent years and is playing an important role in achieving the goal of 100% access to electricity in Rwanda. By June 2022, the proportion of off-grid power supply has reached 22% in Rwanda.

100% access to electricity in the country by 2025, including providing electricity for 35% population through off-grid projects. With the implementation of this Plan, 6 million off-grid accesses will be newly built across the country by 2025 through independent solar energy solutions and microgrid technologies.

3. Kenya

The Kenya Off-Grid Solar Access Project (KOSAP) is a flagship project of the Ministry of Energy of Kenya, which was implemented in 2017 to provide power and clean cooking solutions for remote, low-density and traditionally underserved areas in the country. The project is an important component of Kenya National Electrification Strategy (KNES) and also an important incentive measure for Kenya to realize its vision of 2030. The project includes microgrids for households and public facilities, independent household systems for public facilities, solar water pumps for community facilities, improvement of capacity building, etc. Along with the implementation of the Project, Kenya will build 151 microgrids and provide 250,000 independent solar home systems.

4. Ghana

The Ghanaian Government released the Renewable Energy Master Plan (REMP) in 2019, which incorporates off-grid electrification projects and aims to provide 1000 off-grid communities with off-grid renewable energy power generation schemes. In recent years, the Ghanaian Government vigorously promoted the expansion of the Scaling-up Renewable Energy Program (SREP), aiming to accelerate the development of renewable energy by releasing financing opportunities. Among them, off-grid projects include 55 government-invested renewable microgrids and independent solar energy systems invested by the private sector for 33000 families, 1350 schools, 500 medical centers and 400 communities. With the preparations completed in 2021, the plan will be implemented in 2022.

5. Morocco

Since signing *the Paris Agreement* in 2016, Morocco has been committed to vigorously promoting the development of renewable energy and plans to achieve 50% electricity demand from renewable energy by 2030 and 100% by 2050. The off-grid renewable energy is an important part of this strategy. In 2018, the Office National de l'Electricité et de l'Eau Potable (ONEE) cooperated with a third party to install off-grid solar energy systems for 19438 households in over 1000 villages in Morocco. According to IRENA data, Morocco currently has about 128000 households supplied with power through the

Strong financing capability

Today, the international community is very supportive of small and refined off-grid projects. The World Bank, the African Development Bank and many African countries have launched financing incentives for off-grid projects.

Benefiting people's livelihood

The off-grid system can be flexibly arranged in various places in combination with new energy, can be installed easily and used widely, to meet the power demand of industry, agriculture, commerce, medicine and other fields far away from the large power grid.

What initiatives can promote off–grid in Africa?

The off-grid renewable energy model has a great promotion potential in Africa and plays an important role in meeting the power demand of African society, solving the power problem of the population without access to electricity and promoting the energy transition. African countries are actively formulating national electrification strategies with off-grid development as an important part. The incentive policies adopted by some African countries to promote the off-grid renewable energy model are listed as follows.

1. Nigeria

Nigerian Rural Electrification Authority (REA) used to focus on power grid expansion investment, but now focuses on providing off-grid solutions for rural residents. As a part of the Power Sector Recovery Program (PSRP), the Off-grid Electrification Strategy (OGES) developed by REA aims to provide distributed energy solutions for power supply to households, communities and enterprises. It is planned to build 10000 microgrids by 2023 to provide electricity for 14% of the population; provide reliable power supply for 250000 small- and medium-sized enterprises; provide uninterrupted power supply for federal universities and their teaching hospitals. By 2023, 5 million independent solar energy systems will be deployed for residential buildings and small- and medium-sized enterprises.

2. Ethiopia

Early in 2016, the Ethiopian Government formulated the Rural Electrification Fund (REF) strategy, which aims to achieve power supply in rural off-grid areas and complete 45365 solar home system projects in five phases to provide electricity for about 545 rural health stations and about 370 primary schools and training centers. The Ethiopian Government formulated the National Electrification Plan 2.0 (NEP2.0) in 2019, which aims to achieve

What are the characteristics off–grid?

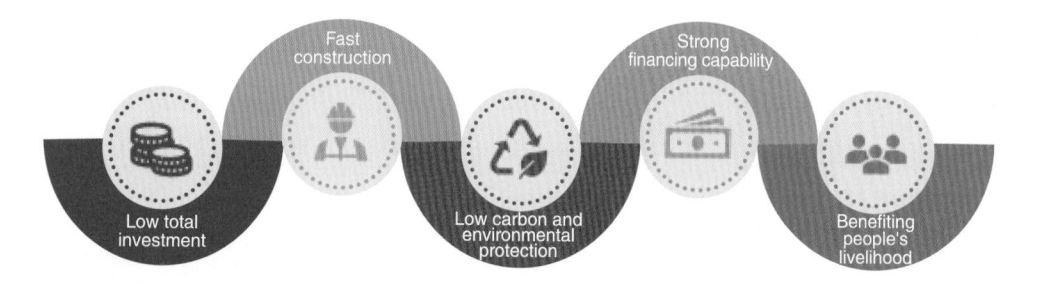

Fast construction

Strong financing capability

Low total investment

Low carbon and environmental protection

Benefiting people's livelihood

Low total investment

The off-grid system is so flexibly that it can be installed nearby at the users. It can make full use of local renewable energy resources and does not need to build large power stations and long-distance power transmission networks. Its total cost is low.

Fast construction

The off-grid is characterized by short construction cycle, immediate efficiency, high replicability, convenient centralized management and less spatial limitation. Its capacity can be flexibly adjusted in light of the demand growth, making combination and expansion very easily.

Low carbon and environmental protection

The carbon emission of off-grid PV power generation accounts for 1/20 to 1/10 of that of fossil energy power generation. With renewable energy as the main power source to achieve independent use, it is a green and low-emission form of power generation.

China Africa Cooperation (FOCAC) points out: "China and Africa will strengthen practical cooperation energy under the framework of the China-AU Energy Partnership to jointly improve electrification in Africa, increase the proportion of clean energy and gradually solve the problem of energy shortage."

1.4 THE IMPORTANCE OF PROMOTING THE OFF-GRID MODEL IN AFRICA

One important factor making it difficult to improve electricity availability in Africa is that over 80% of African people without access to electricity are in remote rural areas with traffic block and dispersed population. Moreover, the power grid has a relatively small coverage in Africa and is even not available in many regions. Therefore, total investment would be too great to rely on the extension of the main grid to provide access to electricity for these people. In Africa, renewable energy is locally accessible, and the off-grid layout is flexible with mature technology. Total cost of construction is much lower than that of the main grid. This is in line with the actual situation in Africa. Therefore, the international community has increasingly attached importance to the off-grid renewable energy model to provide power supply for people without access to electricity.

What is off–grid?

Off-grid is a power generation system, which adopts power supply models with regional independent power generation and household-based independent power generation and operates independently of the main power grid. This power generation system is not subject to geographical restrictions and can be widely used in areas with different conditions. Renewable energy is mostly used for power supply. So, it is very suitable for remote areas without power grids and can also be used as emergency power generation equipment in areas with frequent power outages. It is highly practical for households, enterprises, factories and other users in areas without power grids or with frequent power outages. Moreover, the use of off-grid renewable energy is more economical than diesel power generation. Therefore, the off-grid power generation system has a great potential for application in areas without power grids or with frequent power outages.

G20

The Group of 20 (G20) is one of the largest economic organizations in the world, accounting for 90% of total global economic output. It has regarded energy access as an important issue in the energy field for a long time. Early in 2015, G20 members adopted the *G20 Action Plan on Energy Access: Voluntary Cooperation on Energy Accessibility.* According to the actual situation and development points of each country, they made a voluntary commitment to strengthen cooperation and knowledge and experience sharing to jointly provide electricity to 10% of global population without access to electricity. The first phase focused on improving the electricity consumption in Sub-Saharan Africa.

As COVID-19 led to the first increase in the number of people without access to electricity in the world, the G20 Energy Ministers Meeting issued a communiqué in September 2020, reaffirming the commitment to accelerating energy access, calling on Member States and relevant international organizations to consider accelerating the clean cooking and electrification process voluntarily, and emphasizing the establishment of a results-oriented financing mechanism; strengthening the construction of clean cooking market; assisting the government in formulating comprehensive energy plans which require commitments by all countries; and strengthening the capacity building of public and private sectors in the target countries.

China

Driven by the "Belt and Road Initiative", Chinese enterprises have carried out electricity cooperation in about 70% African countries. From 2010 to 2020, Chinese enterprises participated in the construction of about 150 power plants and power transmission and distribution projects in Africa. More than 100 million people in Africa have gained access to electricity through the power grid, of which the contribution rate of Chinese enterprises has reached 30%. According to the forecast of IEA in 2019, China is expected to complete 49 power generation projects in Africa by 2024, most of which are renewable energy projects, equivalent to 20% of total installed capacity in this region in the same period.

The National Energy Administration (NEA) of China and the African Union Commission (AUC) signed a MoU in 2021 to establish the China-AU Energy Partnership, of which the important content of cooperation is to improve access to electricity in Africa, *The Dakar Action Plan (2022—2024)* adopted at the eighth ministerial conference of the Forum on

Saharan Africa have access to clean, affordable and high-quality off-grid lighting and energy products by 2030. "Lighting Africa" has been supported by the Energy Sector Management Assistance Program (ESMAP), the Public Private Infrastructure Advisory Facility (PPIAF), the Dutch Ministry of Foreign Affairs, the Italian Ministry for the Environment, Land and Sea (IMELS) and the IKEA Foundation. At present, over 32 million African people have gained access to electricity through the project.

In 2019, the World Bank approved the Regional Off-grid Electrification Project (ROGEP), including USD 150 million in credits and grants from the International Development Association (IDA) and USD 74.7 million in emergency recovery grants from the Clean Technology Fund, to help off-grid power supply in 19 West African countries and the Sahel region. The project aims to create a regional market for independent solar energy systems with a regional coordinated approach to increase the power supply of households, enterprises and public institutions in the region and is expected to benefit about 1.7 million people without access to electricity.

African Development Bank

The African Development Bank (AFDB) launched the *New Deal on Energy for Africa in 2016*. By 2020, USD 12 billion had been invested and USD 850 million was mobilized to promote energy access and clean energy transition. The strategy encourages that the distributed energy should be adopted to help 3 million people gain access to electricity through off-grid technologies, e.g. solar home systems. In 2020, Kenya obtained about USD 150 million from the African Development Bank to implement the off-grid electrification plan in Kenya and provide solar energy for 250000 families in 14 counties.

In August 2021, the African Development Bank (AFDB) has reached consensus on a financing agreement for a $20 million concessional investment from the Sustainable Energy Fund for Africa (SEFA) for the COVID-19 Off-Grid Recovery Platform (CRP). The platform supported enterprises to commercialize solar home systems, green microgrids, clean cooking and other renewable energy solutions to mitigate the impact of the epidemic and promote the recovery of the industry. This five-year financing initiative with a total amount of USD 50 million aims to provide relief and recovery funds for areas accessible to energy.

♦ The Sustainable Development Solutions Network (SDSN) research shows that 54 African countries face challenges in achieving the Sustainable Development Goal 7 (SDG7) and over 75% of these countries still face "major challenges". Among them, only Gabon is expected to achieve the SDG7, 3 countries are "declining", 28 countries are "stagnant", and the remaining 22 countries are only "moderately growing".

♦ Mobilizing the enthusiasm of financial institutions and donors is crucial to reducing the imbalance in regional development in Africa and promoting the process of electricity availability in Africa. According to IEA's research, it needs to invest about USD 25 billion every year in light of the current progress in order to achieve the UN's electricity availability goal by 2030, which is about one quarter of total energy investment in Africa before the pandemic outbreak.

1.3   THE WORLD IS MOVING TO IMPROVE ACCESS TO ELECTRICITY IN AFRICA

The urgent task for Africa and international communities is to provide affordable modern energy for all Africans. USD 25 billion every year should be invested in order to achieve this goal by 2030. It is morally unacceptable that persistent injustice of energy poverty in Africa cannot be solved to the best of our ability.

—— Fatih Birol, Executive Director of IEA

Promoting access to electricity in Africa is the long-term vision of the international community. In order to solve the problem that economic and social development has been constrained by electricity shortage in a long term and improve people's well-being pragmatically, countries around the world have been committed to accelerating the development of renewable energy in Africa. They hope to make use of rich local renewable energy resources to ensure that all African families, factories and enterprises can have modern, efficient, reliable, economic and low-carbon clean energy by building energy infrastructure, so as to create more employment opportunities and promote all-round economic and social development.

World Bank

The World Bank launched "Lighting Africa" in 2007, which is one of the important contributions to the realization of "Sustainable Energy For All". It mainly aims at African countries and plans to make 250 million people without access to electricity in Sub-

Africa has about 630 million people without access to electricity, mainly in Sub-Saharan areas. The overall rate of electricity availability in Africa is 58% with large differences from one region to another. The rate is high in the north but low in the south. The rate has reached 98% in North Africa, but there is a long way for other regions to achieve the goal of electricity access. Therefore, it is required to energetically provide reliable and extensive power supply for Sub-Saharan Africa in order to help Africa accelerate the progress of electricity availability.

Fig. 1-3   Electricity shortages in various regions of Africa

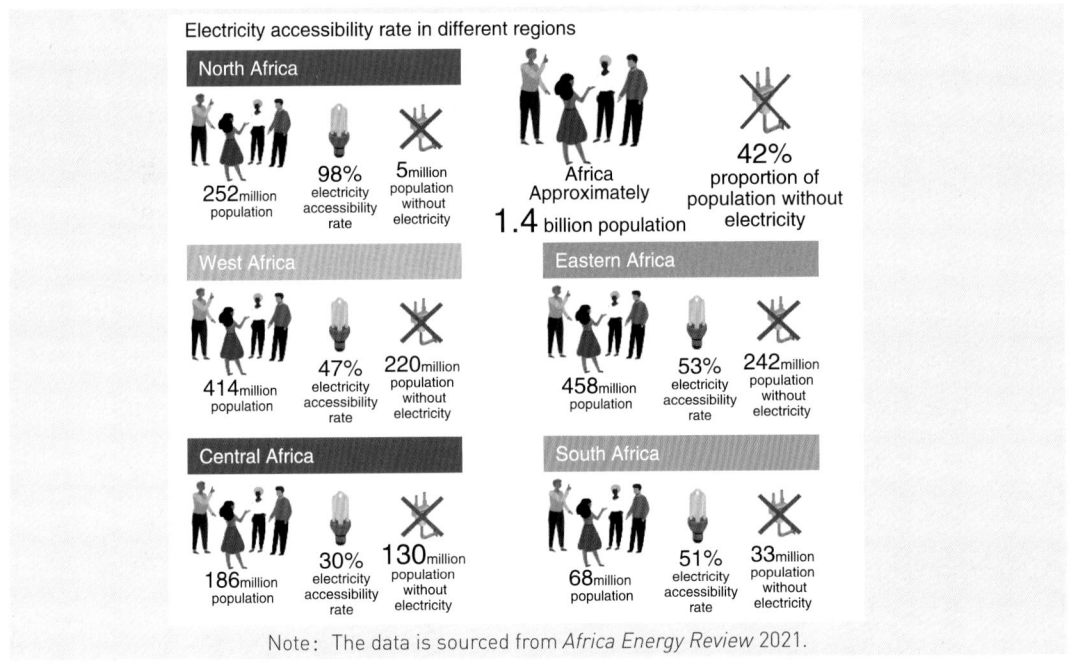

Note：The data is sourced from *Africa Energy Review* 2021.

◆ The IEA forecasts that about 8% of global population will still have no access to electricity by 2030, namely, about 670 million people will have no access to electricity services. It is necessary to accelerate the process of achieving electricity availability to less developed countries, as well as fragile and conflict-affected countries, in Africa.

◆ The rate of access to electricity increased from 49% in 2010 to 58% in 2020, showing positive progress. However, it is worth noting that most people with access to electricity in the past 10 years are in North Africa. In comparison, the rate of electricity availability is rising slowly in Sub-Saharan Africa, pushing the most disadvantageous people more deeply into backwardness.

◆ The Africa *Energy Outlook 2022* released by the IEA shows that the number of people without access to electricity increased by 13 million in Africa in 2020 owing to the COVID-19. It was the first time that Africa deviated from the goal of eliminating energy poverty for 600 million people since 2013 and ended the progressive trend of electricity availability in the past 6 years.

Access to electricity constrains Sustainable development in Africa

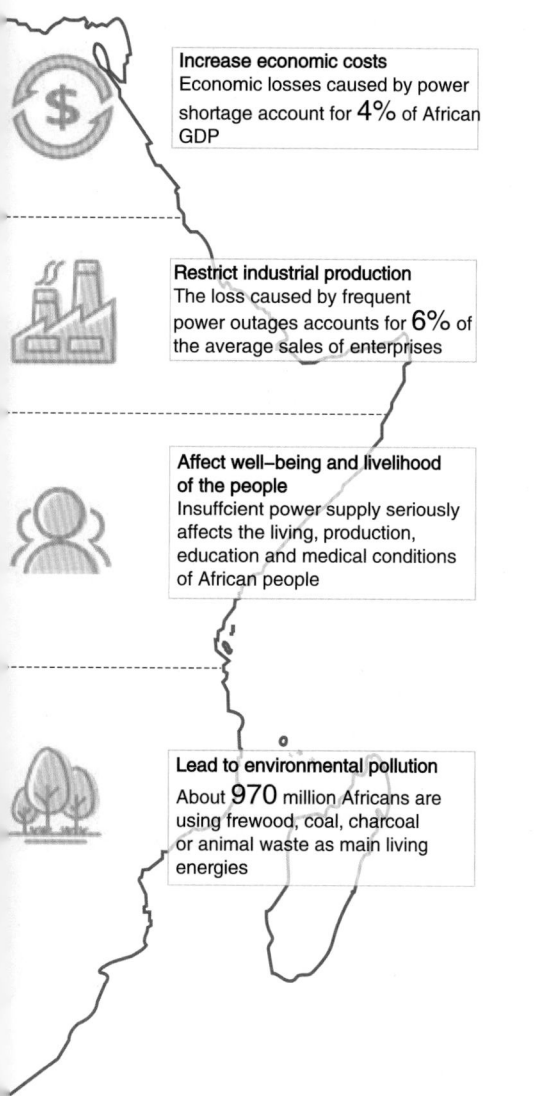

Increase economic costs
Economic losses caused by power shortage account for **4%** of African GDP

Restrict industrial production
The loss caused by frequent power outages accounts for **6%** of the average sales of enterprises

Affect well–being and livelihood of the people
Insuffcient power supply seriously affects the living, production, education and medical conditions of African people

Lead to environmental pollution
About **970** million Africans are using frewood, coal, charcoal or animal waste as main living energies

With 13% of global population, Sub-Saharan Africa has accounts for 4% of energy consumption, 1/3 of per capita energy consumption and less than 3% installed power generation capacity in the world. A large number of people without access to electricity (accounting for 75% of the world's total without access to electricity) and power shortage are still the primary problems restricting Africa's socio-economic development. Low incomes and inefficient and expensive forms of energy supply make energy affordability a key challenge.

♦ More than 30 African countries are suffering from power shortages and frequent service interruptions: 25 days in Senegal, 63 days in Tanzania, 144 days in Burundi on average every year...

♦ Although Nigeria is the largest member of the Organization of Petroleum Exporting Countries (OPEC) in Africa, local enterprises have lost about USD 29 billion every year due to unstable power supply.

♦ In July 2022, South Africa was implementing the second Level 6 power rationing in its history. The power outage lasting over 6 hours a day has seriously affected normal production and life of South Africans.

♦ In Africa, the use of firewood and charcoal has released a large number of pollutants and carbon dioxide into the air and the economic losses caused by air pollution are up to hundreds of billions of US dollars every year.

1.2   AFRICA IS A REGION WITH THE MOST SEVERE ELECTRICITY SHORTAGE IN THE WORLD

AFRICA HAS A GREAT ECONOMIC AND SOCIAL DEVELOPMENT DIVIDEND

Africa, the second largest continent in the world, plays an important role on the global stage. It is a hot land with unlimited possibilities. It is a vast territory with rich natural resources and favorable land and energy. It plays a very important strategic role in the global political and economic pattern. Over the past two decades, African economy has been growing continuously and Africa is expected to be one of the fastest growing regions in the world.

♦ Africa has vast and fertile land, covering a total area of 30.2 million km², where there are vast available land resources and 75% of the area are plateaus and plains.

♦ With a total population of 1.28 billion, Africa has a huge demographic dividend, ranking second only to Asia, and over 50% of the population is young people under the age of 20.

♦ In Africa, there are rich clean energy resources. The world's longest river, the Nile River, and the world's second largest river system, the Congo River, lie in Africa. Moreover, there are rich wind and solar energy resources. Water, wind and solar energy resources account for 12%, 32% and 40% of total global resources, respectively.

♦ The World Bank has committed USD 5 billion to new technology and financial support for energy projects in six African countries. The Chinese Government regards electricity investment in Africa as an important and priority cooperation topic in the "Belt and Road Initiative".

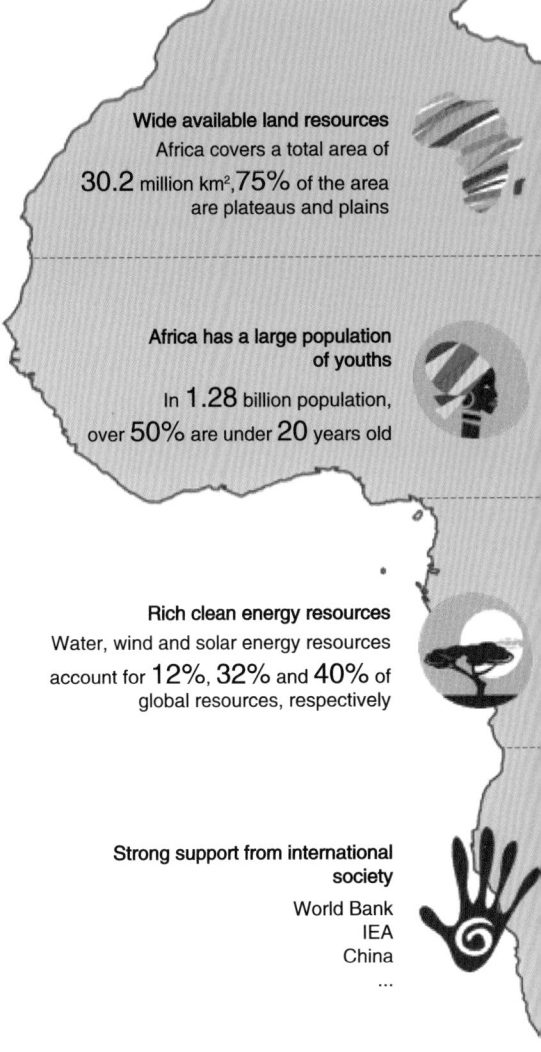

Wide available land resources
Africa covers a total area of
30.2 million km², 75% of the area
are plateaus and plains

Africa has a large population of youths
In 1.28 billion population,
over 50% are under 20 years old

Rich clean energy resources
Water, wind and solar energy resources
account for 12%, 32% and 40% of
global resources, respectively

Strong support from international society
World Bank
IEA
China
...

At present, the annual per capita electricity consumption is about 200 kWh in African countries, which is less than that of a refrigerator and is insignificant while compared with the annual per capita electricity consumption of 6500 kWh in the Europe and 13000 kWh in the United States. Lack of infrastructure, insufficient power investment, low power grid coverage, low new energy development rate, over-dispersed population, weak technology and poor management level are main reasons for access to electricity. At present, insufficient power supply has become a key constraint to Africa's economic and social development.

1.2 AFRICA IS A REGION WITH THE MOST SEVERE ELECTRICITY SHORTAGE IN THE WORLD

Africa cannot develop in the dark.

——Thomas Kwesi Kwad, Vice President of the African Union

Many children in Africa do not yet have access to electric lights to read at night

Many households in Africa cannot yet use electricity for clean cooking

Many shops in Africa are not equipped with necessary electrification equipment

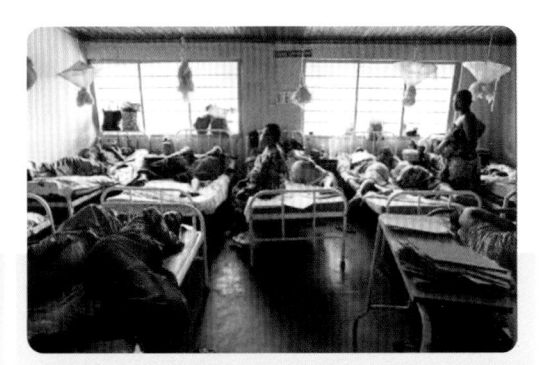

Many hospitals in Africa still lack electrified medical facilities

Electricity shortage is very prominent in Africa. There are about 730 million people without access to electricity in the world, three quarters of whom live in Sub-Saharan Africa. In this region, many African countries will still be unable to access the power infrastructure to meet their energy needs for a long time to come. By 2021, about 630 million people still lacked power supply in Africa, accounting for 42% of total population in Africa.

and farmers to use clean irrigation systems and processing machines for higher outputs and incomes.

♦ In Id Mjahdi, a remote mountain village in Morocco, the access to solar energy has enabled children to read at night , people to save the trouble of getting water from wells several kilometers away and the income of local families to increase by bringing employment opportunities of making nut oil and other products.

♦ The Global Off-Grid Lighting Association (GOGLA) has surveyed 2300 households in East Africa, among which about 58% have installed off-grid solar energy systems and about 36% have increased their monthly income by an average of USD 35, equivalent to 50% of the average monthly GDP of households in this region.

♦ Due to the lack of reliable power supply, some countries in Sub-Saharan Africa may lose nearly 7% of their GDP every year and 30% of medical facilities lack power, undermining the medical conditions for 255 million people.

♦ At present, about 1 of every 10 people live without access to electricity in the world. Consequently, they have not access to opportunities for working, studying or running a business brought by electricity, or more safe and secure medical services.

♦ Due to the lack of access to electricity and clean fuel, about 2.4 billion people in the world are still using firewood, coal, dung and other ways for heating and cooking, so that many families are beset by smoke and fire and suffer heart disease, stroke, cancer, pneumonia and other diseases, leading to millions of deaths every year.

1.1 ACCESS TO ELECTRICITY IS AN IMPORTANT ISSUE OF GLOBAL CONCERN

Continued

Country		Per capita electricity consumption /(kWh/a)	Proportion of electricity access / %	Per capita GDP / (USD/a)
South Africa region	Comorin	46	70	1362
	Lesotho	430	46	1003
	Madagascar	59	27	502
	Malawi	62	11	407
	Mauritius	1976	99	8993
	Mozambique	384	38	450
	Namibia	1479	45	4175
	Seychelles	3391	99	11639
	Swaziland	1296	76	3054
	Tanzania	97	38	1090
	Zambia	634	33	981
	Zimbabwe	489	53	1385
	South Africa	4064	99	5059

WHY IS ACCESS TO ELECTRICITY SO IMPORTANT?

◆ According to the data of Rockefeller Foundation, the access to electricity can increase per capita household incomes by 39%, enable enterprises to operate at a higher level

Continued

Country		Per capita electricity consumption /（kWh/a）	Proportion of electricity access / %	Per capita GDP / （USD/a）
East Africa region	Ethiopia	84	48	994
	Kenya	147	78	2039
	Rwanda	41	55	819
	Somalia	27	35	327
	South Sudan	34	7	296
	Sudan	266	47	775
	Uganda	72	26	912
West Africa region	Nigeria	115	68.2	2083
	Benin	89	31	1251
	Côte D'Ivoire	274	78	2278
	Ghana	319	85	2223
	Senegal	222	70	1460
	Togo	146	46	905
	Burkina Faso	74	21	791
	Cape Verde	630	95	3148
	Gambia	130	62	791
	Guinea	44	44	1106
	Guinea-Bissau	193	9	790
	Libya	55	29	3281
	Mali	153	52	897
	Mauritania	264	47	1971
	Niger	47	14	566
	Sao Tome and Principe	291	77	1918
	Sierra Leone	42	22	527
South Africa region	Angola	278	45	2012
	Botswana	1569	60	6781

1.1 ACCESS TO ELECTRICITY IS AN IMPORTANT ISSUE OF GLOBAL CONCERN

There are 54 sovereign countries in Africa with main high power consumption region in its northern and southern parts, where the power consumption accounts for about 80% of Africa's total and the annual per capita electricity consumption is about 2000 kWh. The annual per capita electricity consumption is about 200 kWh in other African regions, less than 1/30 of that in Europe and 1/60 of that in the United States, and the annual per capita electricity consumption is even less than 100 kWh in many African countries in these regions. The statistics of per capita electricity consumption and GDP of the 54 African countries are shown in Table 1-1. From these data, it is easy to find that 28 African countries have less than 50% of access to electricity, accounting for over 1/2 of all the sovereign countries. All of these countries are located in the Sub-Saharan region, where the lack of power supply has greatly affected local economic development.

Table 1-1   Africa's per capita electricity consumption and GDP

	Country	Per capita electricity consumption /(kWh/a)	Proportion of electricity access / %	Per capita GDP / (USD/a)
North Africa region	Algeria	1302	99	3263
	Egypt	1534	99	3587
	Libya	3962	99	3281
	Morocco	794	99	3158
	Tunisia	1303	99	3323
Central Africa region	Cameroon	231	63	1470
	Central African Republic	27	5	490
	Chad	12	8	654
	Congo	172	48	2186
	Democratic Republic of the Congo	73	9	541
	Equatorial Guinea	556	67	6773
	Gabon	928	91	7421
East Africa region	Burundi	32	10	254
	Djibouti	409	42	3074
	Eritrea	585	10	588

1.1 ACCESS TO ELECTRICITY IS AN IMPORTANT ISSUE OF GLOBAL CONCERN

It is a double task facing all countries to address access to electricity and climate change.

—— António Guterres, Secretary General of the United Nations

As of 2021, 730 million people were still living without access to electricity, accounting for about 10% of global population. Judged by the current development trend, we are still facing serious challenges to achieve the Sustainable Development Goal 7 by 2030. There are still many regions without access to electricity in the world. Women and children must spend several hours getting water. Clinics cannot store children's vaccines. Many students cannot do their homework in the evening. If the international community could not take immediate actions, it is estimated that 650 million people would still have no access to electricity across the world by 2030. This means that it is unlikely to achieve the UN's goal of ensuring affordable, reliable and sustainable modern energy for all by 2030. At present, further improving the utilization ratio of renewable energy is critical to achieving access to electricity.

Fig. 1-2 Distribution of global population without access to electricity in 2000–2030

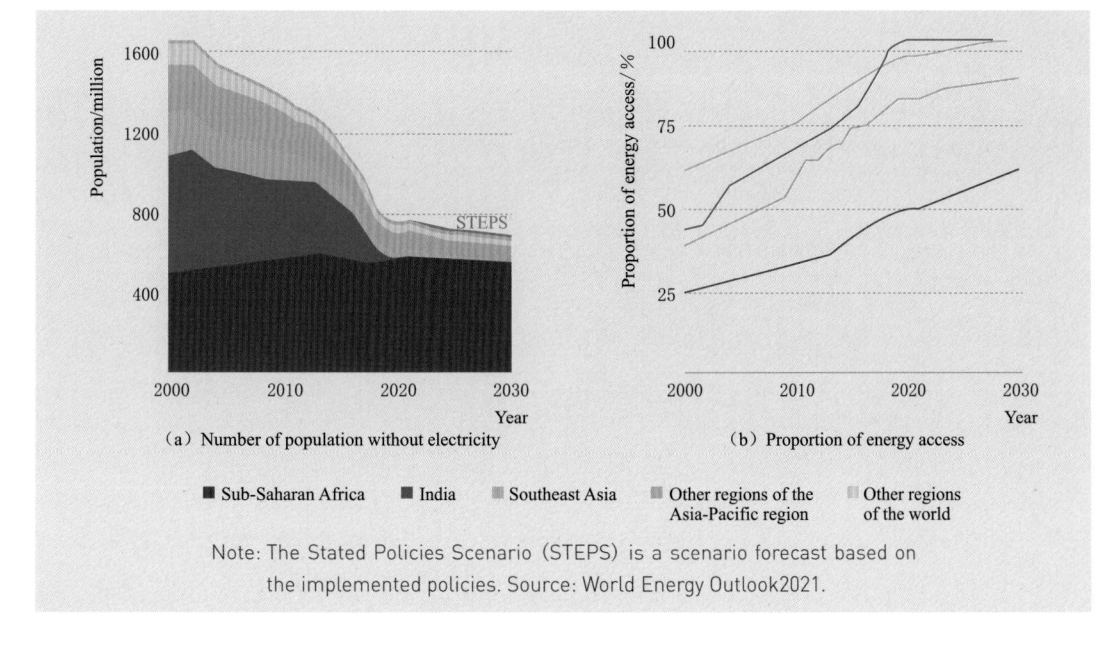

BACKGROUND

Goal 7.1: Ensure access to affordable, reliable and sustainable modern energy for all by 2030.

Indicator 7.1.1: Proportion of urban/rural population with access to electricity (%).

——Sustainable Development Goal 7 of the United Nations

Fig. 1-1 Relationship between electricity and human's production and living activities

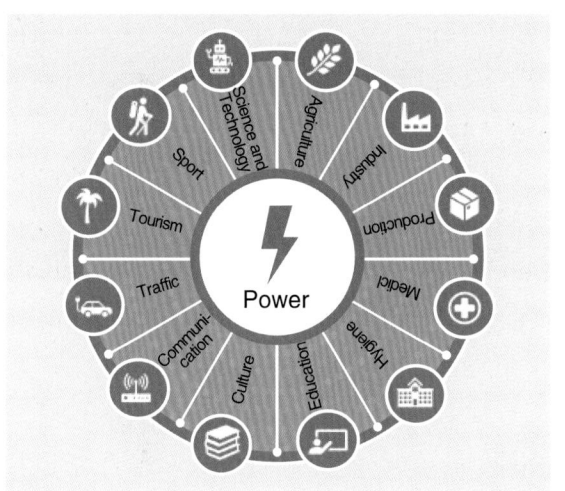

The United Nations regards ensuring access to affordable, reliable and sustainable modern energy for all as one of the important goals of the sustainable development of human society, and the proportion of the population with access to electricity as an important indicator of the extent to which this goal has been achieved. The consumption level of electricity, that is, the extent of access to electricity, has become an important symbol of indicating a country's modernization process and people's living standards. It is difficult to imagine how today's world would work without modern energy services such as electricity.

Electricity is a driving force of modern social and economic operation and one of fundamental energy resources that support the progress of human civilization and economic and social development. The development of electricity is not only vital to major strategic issues such as national economic security, but also closely related to people's daily life and social stability. Electricity has almost become an indispensable and important guarantee in all fields of modern society, ranging from agriculture, industry and production to healthcare, education and communication. Achieving access to electricity is a necessary prerequisite for alleviating poverty, improving social productivity and promoting common prosperity.

1

ANALYSIS ON ELECTRICITY ACCESS IN AFRICA

Africa owns a huge economic development dividend, but the access to electricity problem has constrained Africa's sustainable development for a long time. This chapter analyzes the importance of ensuring electricity consumption for people without access to electricity in Africa under the background of the international community, which is attaching great importance to access to electricity, sorts out the African access to electricity status quo and challenges, and demonstrates the important practical role of the model of off-grid renewable energy renewable energy in ensuring African access to electricity.

ABBREVIATIONS	
STEPS	Stated Policies Scenario
GOGLA	Global Off-Grid Lighting Association
GDP	Gross Domestic Product
SDSN	Sustainable Development Solutions Network
SDG7	Sustainable Development Goal 7
ESMAP	Energy Sector Management Assistance Program
PPIAF	Public Private Infrastructure Advisory Facility
IMELS	Italian Ministry for the Environment Land and Sea
ROGEP	Regional Off-Grid Electrification Project
IDA	International Development Association
SEFA	Sustainable Energy Fund for Africa
CRP	Covid-19 Off-Grid Recovery Platform
IEA	International Energy Agency
REA	Rural Electrification Agency
OGES	Off-grid Electrification Strategy
PSRP	Power Sector Recovery Programme
KOSAP	Kenya Off-Grid Solar Access Project
KNES	Kenya National Electrification Strategy
REMP	Reneable Energy Master Plan
SREP	Scaling-up Renewable Energy Programme
ONEE	National Office of Electricity and Drinking Water
IRENA	International Renewable Energy Agency
ROGEA	Regulation for Off-Grid Energy Access
MEFA	Mozambique Energy for All
FUNAE	Mozambique National Energy Fund
SHS	solar home system
NESPA	Niger Solar Electricity Access Project
NST1	National Strategy for Transformation
AFDB	African Development Bank
LCOE	Levelized Cost of Energy

LISTS OF FIGURES AND TABLES

LIST OF FIGURES

Fig. 1-1   Relationship between electricity and human's production and living activities.......2

Fig. 1-2   Distribution of global population without access to electricity in 2000–2030........3

Fig. 1-3   Electricity shortages in various regions of Africa...12

Fig. 2-1   Flowchart for determination of development models..24

Fig. 3-1   Schematic diagram of geographical location of the agricultural park.................31

Fig. 3-2   Development positioning of the agricultural park...31

Fig. 3-3   Schematic diagram of functional partition of the agricultural park......................32

Fig. 3-4   Schematic diagram of energy consumption of the agricultural park...................33

Fig. 3-5   Four advantages of application of "new energy + agriculture" in Africa...............34

Fig. 3-6   Common application scenarios of "new energy + agriculture" in China.............35

Fig. 3-7   PV energy storage off-grid power generation system.......................................41

LIST OF TABLES

Table 1-1   Africa's per capita electricity consumption and GDP 4

Table 2-1   Division of off-grid user groups 23

Table 2-2   Classification of off-grid development models 24

Table 3-1   Near-term and long-term scenarios of off-grid power supply transformation ... 42

Table 3-2   Equipment electricity consumption statistics for near-term scenarios............ 43

Table 3-3   New equipment electricity consumption statistics for long-term scenarios...... 44

Table 3-4   Configuration of PV power generation system 44

Table 3-5   Configuration of combined cooling, heating and power system 46

Table 3-6   Financial calculation for near-term scheme 49

Table 3-7   Financial calculation for long-term scheme 49

Table 3-8   Sensitivity analysis of project financial calculation................................... 50

Table 4-1   Classification of electricity consumption scenario................................... 54

Table 4-2   Calculation of investment in renewable energy off-grid projects under three

scenarios 56

CONTENTS

PREFACE

FOREWORD

1 ANALYSIS ON ELECTRICITY ACCESS IN AFRICA.. 1

1.1 ACCESS TO ELECTRICITY IS AN IMPORTANT ISSUE OF GLOBAL CONCERN 3

1.2 AFRICA IS A REGION WITH THE MOST SEVERE ELECTRICITY SHORTAGE IN THE WORLD 8

1.3 THE WORLD IS MOVING TO IMPROVE ACCESS TO ELECTRICITY IN AFRICA 13

1.4 THE IMPORTANCE OF PROMOTING THE OFF-GRID MODEL IN AFRICA 16

2 ANALYSIS ON BUSINESS MODEL OF OFF-GRID PROJECTS IN AFRICA ... 21

2.1 DIVISION OF USER GROUPS FOR OFF-GRID PROJECTS 22

2.2 CLASSIFICATION OF DEVELOPMENT MODELS OF OFF-GRID PROJECTS 23

2.3 JUDGMENT OF DEVELOPMENT MODELS OF OFF-GRID PROJECTS..................... 24

2.4 STRATEGIES FOR OFF-GRID PROJECT DEVELOPMENT 26

3 CASE STUDY OF OFF-GRID PROJECTS IN AFRICA 29

3.1 PROJECT OVERVIEW—AGRICULTURAL PARK IN ABUJA 30

3.2 CONCEPTUAL DESIGN OF OFF-GRID TRANSFORMATION OF THE AGRICULTURAL PARK 40

3.3 COMPREHENSIVE BENEFITS OF OFF-GRID TRANSFORMATION OF THE AGRICULTURAL PARK 51

4 DEMAND ANALYSIS AND PROMOTION OF OFF-GRID MODEL 53

5 CONCLUSIONS AND SUGGESTIONS.. 57

5.1 CONCLUSIONS... 58

5.2 SUGGESTIONS 60

STATEMENT 63

sufficiency in covered areas. It is worth noting that in the case of the African region, the population without access to electricity is mostly distributed in remote areas. The electricity affordability of local residents is generally weak, and it is often difficult to support the economical and sustainable operation of off-grid projects only by relying on local residents' electricity revenue. Therefore, in order to facilitate the implementation and promotion of renewable energy off-grid projects in Africa, it is required to carry out in-depth research on business models.

Based on the status quo of African electricity access, this study aims to analyze the classification, judgment anddevelopment strategies of the business model of renewable energy off-grid projects, so as to provide technical reference for economical and sustainable operation of off-grid projects. Meanwhile, taking the CGCOC Agricultural High-tech Industrial Park in Abuja, Nigeria as the study object,through conducting the case study, that is, combining with the practical experience of China's "New Energy + Agriculture" pattern, the conceptual design of renewable energy off-grid upgrading and transformation for the agricultural park is carried out, to analyze and research economic and social benefits of off-grid projects.

FOREWORD

Energy is a common concern of the world today and is at the core of almost every major challenge and opportunity

——— United Nations

As one of the most important energy resources in modern society, electricity has been highly integrated into all aspects of people's lives. Promoting electricity access is an issue of common concern of the international community. In 2021, there were about 730 million people without access to electricity, 75% of which live in Sub-Saharan Africa. It is not difficult to see that Africa is still the most prominent region facing the electricity access problem in the world. China's National Energy Administration (NEA) and African Union Commission (AUC) signed a memorandum of understanding in 2021 to launch the establishment of the China-AU Energy Partnership, which will gradually solve the problem of electricity access in Africa as an important part of practical cooperation between the two sides.

Why is the problem of electricity access in Africa prominent and difficult to solve? The majority of the African population without access to electricity is poor and scattered in vast rural areas that are remote and have traffic block, causing great construction inconvenience and economic costs to the construction of the power grid. Therefore, it is difficult for the traditional extension of the main grid to effectively solve the problem of electricity access in Africa. In recent years, with the continuous progress of renewable energy technology, off-grid power generating systems adopting independent power supply of renewable energy, especially PV and wind power, instead of relying on the main grid, provide new solutions to help eliminate energy poverty in Africa and achieve universal access to electricity for people in remote areas.

Renewable energy has the advantages of cleanliness, low carbon, recyclability, on-site availability, and flexible layout on the user side. Therefore, the off-grid generating system of renewable energy can not only effectively solve the electricity consumption for people without access to electricity in remote areas of Africa, but also achieve electricity self-addressing climate change, promoting energy transformation and improving access to electricity in Sub- Saharan Africa. This important topic was written into the *G20 Energy Access Action Plan: Voluntary Collaboration on Energy Access.* China, together with the international community, has long been committed to helping eliminate energy poverty and improve access to electricity in Africa. In October 2021, the China-African Union Energy Partnership was established. One month later, the *Dakar Action Plan (2022-2024)*, which was adopted at the 8th Ministerial Conference of the Forum on China-Africa Cooperation (FOCAC), pointed out: "China will work with Africa to enhance practical cooperation in the energy sector under the framework of the China-African Union Energy Partnership, jointly improve the level of electrification in Africa, increase the share of clean energy, address step by step the issue of energy accessibility, and promote sustainable energy development of both sides."

Along with accelerated efforts to address climate change and meet carbon neutrality goals, countries around the world are pushing ahead with energy transformation and renewable energy development. Africa boasts excellent renewable energy resources. Renewable energy is clean, low-carbon, recyclable and locally available. Therefore, off-grid power systems that use renewable energy can be flexibly deployed in remote areas of Africa, providing a new solution to power availability for African people who lack access to electricity when the technology of off-grid renewable energy continues to develop and the construction cost keeps declining.

Amb. Rahamtalla M. Osman
Permanent Representative of the African Union to China

PREFACE

Electricity constitutes an important material basis for the progress of human society and acts as a key factor for social and economic development today. The electricity demand is on the increase in the wake of the scientific & technological development and new technologies emerging one after another. In this context, ensuring power supply has become a social and political issue that matters the national energy security, economic and social development and people's well-being, rather than an economic issue alone. In 2021, the global economic growth and more extreme weather conditions spurred the electricity demand to soar by more than 6% worldwide. Countries around the world are speeding up power supply, but multiple problems have remained for residents in areas without access to electricity.

Providing 100% access to electricity through the constant improvement of power supply is a long-term goal of the international community. The Sustainable Development Goals 7 proposes to "ensure access to affordable, reliable, sustainable and modern energy for all by 2030". No one will be left behind. The *African Union's Agenda 2063* urges to develop renewable energy and increase the penetration rate of electricity. This is a long-term vision adopted by African state leaders through the African Union, which is highly consistent with Africa's expectations for prosperity and inclusive growth.

There were still 730 million people around the world living without access to electricity by 2021. Of them, 3/4 are residents in Sub-Saharan Africa. Africa continues to be the most prominent region in the world for access to electricity. The per-capita energy consumption in Africa is only 180 kWh per year, far below the average in European and American countries, which is more than thousands, even tens of thousands kWh per year. Surveys by relevant institutions show that economic losses brought by the electricity shortage to Africa accounts for about 4% of the local GDP. The electricity shortage has undermined living, production, education and medical conditions for African people and become a major constraint to the economic and social development of Africa.

Improving access to electricity in Africa is a topic of common concern for the international community. As early as 2015, the G20 Leaders' Summit focused on

EDITORIAL COMMITTEE

Director:	Li Sheng	Yi Yuechun		
Deputy Director:	Gu Hongbin	Zhang Yiguo	Yu Bo	Gong Heping

Editor in Chief:	Jiang Hao	Chen Zhang	
Associate Editor:	Yang Xiaoyu	Wang Yuliang	Wang Xianzheng
	Wang Jun	Wang Peiyuan	Zhou Gang
	Shuai Dong	Li Yanjie	Xu Xiaoyu
	Xia Yucong	Liu Zichu	Jing Heran
	Huang Jin		
Consulting:	Miao Hong	Song Jing	Teng Aihua
	Zhao Yaohua	Wang Miao	Wang Jun

ANALYSIS ON ELECTRICITY ACCESS IN AFRICA AND CASE STUDY OF OFF-GRID RENEWABLE ENERGY

China Renewable Energy Engineering Institute

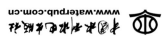

www.waterpub.com.cn

· 北京 ·